GUNPOWDER

GUNPOWDER

A History of the Explosive That Changed the World

JACK KELLY

Atlantic Books
London

First published in the United States of America by Basic Books, a member of the Perseus Books Group.

First published in Great Britain in hardback in 2004 by Atlantic Books, an imprint of Grove Atlantic Ltd

1 2 3 4 5 6 7 8 9

A CIP catalogue record for this book is available from the British Library.

ISBN 1 84354 190 4

Printed in Great Britain by Creative Print & Design, Ebbw Vale, Wales

Atlantic Books
An imprint of Grove Atlantic Ltd
Ormond House
26–27 Boswell Street
London WC1N 3JZ

CONTENTS

PROLOGUE:
THE DEVIL'S DISTILLATE

"... that puissant monarch,
Which rides triumphing in a chariot
On mist-black clouds, mix'd with quenchless fire,
Through uncouth corners in dark paths of death."
— BARNABE BARNES, 1607

FIRE IGNITES OUR dreams and our anxieties. It speaks to us in a language more basic than thought. Our instincts respond to the flicker of flame, to the wavering colors of the coals, to the roar of the conflagration.

Fire needs fuel, oxygen, heat. It needs an initiator—a tiny bit of burning metal struck to white heat by friction against flint, a spark. The heat of the spark rips apart molecules of fuel. Carbon and hydrogen atoms combine with oxygen. The reactions are exothermic—they give off heat to ignite more fuel, a chain reaction. The complex process remains something of a mystery to science even today. We understand roughly what is happening, but the flame appears to have a life of its own. Its energy bursts out as heat, which makes particles of soot incandescent.

Mankind has lived with natural fire for eons—the hearth, the campfire, the candle flame have been our intimates. Like human lungs, the flames are nourished by oxygen from the air. As convection carries away the hot spent gases, fresh air reaches the fuel. But oxygen makes up only 20 percent of the atmosphere. The thirst for oxygen puts a perpetual brake on natural flames. Winds fan a fire—smothered, the flames die.

What if the fire's heat induced oxygen to burst from the very pores of the burning material? The brake would be let off —the fire would burn unrestrained, with utter abandon. The chain reaction of combustion would accelerate at an astonishing rate. Instead of needing minutes or hours to burn, the fuel would go up in a fraction of a second.

This violent reaction, a product of inner oxygen, is man's fire, concocted, singular, unquenchable. It does not exist anywhere in nature. It is "artificial fire"—*feu d'artifice, fuegos artificiales*—terms for what in English we call fireworks, pyrotechnics. Its embodiment is gunpowder.

Artificial fire requires an oxidizer, a chemical that emits oxygen when heated. Mix the oxidizer with the fuel. Grind it until the ingredients are in intimate contact. The oxidizer is saltpeter, the fuel a combination of charcoal and sulfur. As the fuel burns it decomposes the saltpeter, releasing virgin oxygen. The oxygen accelerates the burning, a process technically called a deflagration. You have created gunpowder.

The substance that was to be known as gunpowder was not invented for the gun. Before gunpowder's inception, no one had conceived of a projectile-throwing machine driven by chemical energy. Humans developed tools for using this new material only after it had emerged from the fantastical speculation of alchemists. Only through centuries of trial and error did gunpowder reveal its properties and possibilities.

No rational theory guided the inventors of gunpowder. What's more, during the nine hundred years when powder was in common use, indeed during the century since it has been rendered obsolete for most uses, no other combination of natural ingredients was found that could replicate its effects. Gunpowder was unique.

Early in its history, gunpowder was labeled the "devil's distillate." Onlookers were terrified by its flash and boom. Its fabricators were secretive, blackened men, daredevils whose arcane work was subject to disastrous accidents. One of gunpowder's ingredients, brimstone, was the burning stone always associated with Satan. Gunpowder's action was a diabolical mystery—once ignited, it blazed wildly, infernally, leaving behind the sharp tang of sulfur and a haze of smoke.

Gunpowder, for most of a millennium, was mankind's only explosive. It was one of the few chemical technologies to emerge from the Middle Ages. Its effects were momentous. In the seventeenth century, Francis Bacon spoke of "those three which were unknown to the ancients, and of which the origin, though recent, is obscure and inglorious; namely printing, gunpowder and the magnet. For these three have changed the whole face and state of things throughout the world." Gunpowder was indeed of inglorious origin, fashioned by craftsmen from the basest of ingredients. It was just as surely a catalyst of the modern world, an invention that threw up a divide beyond which the rivers of history flowed in a new direction.

Today, gunpowder is an anachronism. The powdermakers who operate the few remaining mills use methods that are centuries old. Their way of making powder would not be a mystery to an artisan of the 1300s. It is remarkable that a technology that arrived in the West in the time of Dante was still performing valuable service in the time of Henry Ford. A substance that was fueling skyrockets and firecrackers during the era of Genghis Khan will be doing the same during the era of the quantum computer.

This book is about that original technology, about the powder that resulted from the mechanical mixing of naturally occurring ingredients. During the latter part of the 1800s, this ancient substance was superceded by synthetic propellants and explosives derived from the chemistry laboratory. The original powder came to be known as "black powder" to distinguish it from its modern cousins—smokeless powder, cordite, dynamite, TNT. Through most of its history, the substance was referred to simply as "powder."

The principal use remaining for gunpowder is the fabrication of fireworks. A gunpowder charge hurls aloft the shells of the pyrotechnic display. A gunpowder fuse hisses toward the casing as it flies. Another charge bursts the bomb to let loose burning nuggets whose colorful glow creates such splendid effects. The smoke that drifts into the crowd smells the same as the smoke that wafted through ancient China, that saturated innumerable battlefields, that seeped up from coal mines—it is the pungent, evocative smoke of history. The final function of gunpowder is the same as the first. Before flamethrowers, bombs, and guns filled the world with their terror, gunpowder was the servant of delight and the handmaiden of wonder.

1

FIRE DRUG

IN THE MOUNTAINS of western China, legendary semi-human monsters called *shan* peeked through the leaves at the campfires of travelers. When the men were away, the naked creatures snuck close to steal salt and to roast frogs and crabs over the flames. If confronted head-on, the *shan* could afflict their assailants with fever.

The best way to drive off these ogres was to throw bamboo into the fire. The expanding air and steam inside the plant segments created pressure that burst the twigs with a loud snap. In this manner the superstitious travelers created a little explosion, a word whose Latin roots mean to drive out by clapping. All mammals are equipped with a startle reflex, a primitive brain circuit that makes them tense, jump, cringe in response to a loud noise. The Chinese assumed that *shan* reacted the same way.

These early noisemakers were used from prehistoric times and became a common amusement. On New Year's Day in China the crack of exploding bamboo frightened away evil spirits and cleared the way for the coming year. Exploding bamboo was still popular when Marco Polo brought home his wondrous account of Cathay in 1295. Young green canes thrown into a fire "burn with such a dreadful noise that it can be heard ten miles at night," he claimed, "and anyone who was not used to it could easily go into a swoon or even die."

His description reminds us how quiet the medieval world was—no motorized vehicles or amplified music to disturb the sleep. Thunder was the loudest noise anyone had ever heard. Even war was relatively quiet, limited to the pounding of drums, the shouts of combatants, and the clank of weapons. From over the next hill, the cacophony of a battle would dissolve to a whisper.

During the tenth century A.D., a new instrument for making noise came into being—one based on a unique mixture of ingredients. A medieval Chinese text with the fanciful title "Dreams of the Glories of the Eastern Capital" describes a display that the Chinese army mounted for the emperor around A.D. 1110. The spectacle opened with "a noise like thunder" and continued with fireworks exploding against the blackness of the medieval night. Dancers in strange costumes followed, moving through clouds of colored smoke.

The substance that produced these sensational effects would have a singular impact on societies around the world. Yet the material entered history slowly, tentatively, emerging out of centuries of happenstance, observation, trial and error. Only gradually did men understand that they were dealing with something genuinely new on earth. It was a combination of saltpeter, sulfur and charcoal, ground laboriously together in just the right proportions. The Chinese called it *huo yao*, "fire drug."

——

TAOISM WAS THE catalyst for this radical new technology. The system of thought introduced by Lao Tzu in the sixth century B.C. began as a purely philosophical discipline. But a later branch of the tradition absorbed the magical beliefs of earlier folkways and incorporated a mixture of sorcery, superstition, and esoteric knowledge. Taoists' interest in magical manipulations became established as Chinese alchemy.

Alchemy contributed three critical elements to the discovery of "fire drug": purification, observation, and experiment. Chinese alchemists labored to rid of adulterants the substances they found in nature. Purity was a sacred quality, refining a rite. Even a small amount of contamination in "fire drug's" ingredients could derail the combustion reaction.

Alchemists puzzled over how the five basic elements of the natural world—metal, wood, earth, water, and fire—could interact to produce the manifold universe. They took note of quirks, such as a quickness of combustion, that might have otherwise escaped attention. Having observed, they experimented. While not scientific in the modern sense, their systematic trial and error enabled them to grope into the unknown.

In the West, alchemy focused on ways of transforming base substances into gold. In contrast, the principal aim of Chinese practitioners was to create an elixir of immortality. Their interest was drawn by materials with paradoxical properties—gold, the element that never tarnished; mercury, the liquid metal; sulfur, the stone that burned.

Was it possible that these materials held the secret of perpetual youth? Taoist alchemists conducted a centuries-long search to find the proper combination. The emperors themselves were susceptible to the allure of their potions. The capable Tang dynasty emperor Li Chun, who ruled from A.D. 806 to 820, was one of several who fell under the influence of the alchemists. Hoping to live forever, Li became an inveterate consumer of elixirs.

The problem with the nostrums was that many of their exotic ingredients—like white lead and the chemical realgar or arsenic sulfide—were deadly poisons. The ingestion of mercury, a favorite ingredient, can cause ulceration of the gums, fever, bloody vomiting, burning pains, and muscle tremors. It also affects the mind, distorting the senses and giving rise to melancholia and mania.

Courtiers must have been disconcerted to observe the emperor, the pole star around whom the entire world revolved, go mad. This was Li Chun's fate. He was warned by his postal service minister that the alchemists "have come for nothing but profit." The minister was demoted; the emperor persisted in his folly. His behavior became increasingly erratic until he was assassinated by court eunuchs.

The elixirs held even more portentous dangers. A book dating from around A.D. 850 called "Classified Essentials of the Mysterious Tao of the True Origin of Things" debunks thirty-five elixirs. Of one it warns, "Some have heated together sulfur, realgar and saltpeter with honey; smoke and flames result, so that their hands and faces have been burnt, and even the whole house where they were working burned down."

This casual warning marks an epoch in human history. It was the debut of artificial fire on earth, the whispered beginning of the long and momentous history of gunpowder. Alchemists had stumbled across a clue to the magical effects of which saltpeter was capable when mixed with sulfur and a source of carbon—in this case dried honey. They had created a material with a startling new relationship to fire. It was not yet true "fire drug," the explosive that would become known around the world as gunpowder. But the observant and inquisitive alchemists, searching for the key to immortality, had taken an important step in a dramatic new direction.

———

SALTPETER WOULD become the central component of "fire drug." Otherwise known as niter, the salt was readily available in China as a white crust on certain soils. Alchemists had studied its qualities for centuries. They mixed it with water to form a weak nitric acid solution with which they could dissolve otherwise insoluble materials. In areas where table salt was in short supply, cooks sometimes used saltpeter to enhance the flavor of food. They no doubt noticed that when a pinch was thrown into the fire it caused the flames to flare up. All the alchemy manuals from the Tang dynasty (A.D. 618–907) mention saltpeter and its preparation. The production of relatively pure saltpeter was part of the alchemist's repertoire.

Saltpeter is a waste product of two strains of bacteria that are among the many that feast on decaying organic matter. These enterprising microorganisms—*Nitrosomonas* and *Nitrobacter*—are friends to the organic gardener, transforming the raw materials of rot into the nitrates plants love.

In the nitrate radical that is the crux of saltpeter, three oxygen atoms are fastened to one of nitrogen. This unit forms salts with whatever metals are available, such as calcium, sodium, or potassium—the last being most valuable for making gunpowder.

Nitrates are among the most soluble of all salts. Saltpeter dissolves in rainwater, soaks into the earth, and is wicked upward by evaporation. All other materials solidify first, leaving the nitrates to concentrate at the surface. Southern China's hot climate, accompanied by alternating rainy and dry seasons, promoted both rapid decay and speedy evaporation. In some places the conditions yielded crude saltpeter that could be extracted from the top layer of the soil.

What made saltpeter the essence of "fire drug" was the property already observed by Chinese cooks. When exposed to heat of 335° centigrade, the normally stable salt breaks down, letting loose the oxygen atoms that had been bound up with nitrogen. The key mechanism of artificial fire was this release of virgin oxygen, which became available to burn any surrounding fuel. How much saltpeter to include in the mixture was a matter that could only be worked out through long trial and error. Three-quarters of the total by weight would eventually emerge as the ideal proportion—early "fire drug" containing lesser amounts burned vigorously but did not explode.

All that was needed to complete the formula was a readily available fuel. The flare-up of the dangerous elixir may have pointed the way. Alchemists had long known of sulfur's fiery potential. One of the few elements to exist in a pure state in nature, sulfur could be found in deposits near volcanoes, or it could be obtained by heating pyrite ores and allowing the sulfurous vapors to solidify on the cool walls of a container.

The other fuel added to the mix was charcoal, which had long been used as a source of heat. The remains of wood that has been cooked in an oxygen-poor environment, charcoal is a complex ingredient that consists of pure carbon laced with volatile hydrocarbons and other remnants of its organic source. These chemicals, along with charcoal's latticelike structure, play a subtle but crucial role in the action of gunpowder that scientists to this day do not completely understand. Powdered coal and graphite, while almost pure carbon, lack these additions and will not make effective gunpowder.

Saltpeter, sulfur and charcoal rely on a unique and intricate teamwork to bring about the magic associated with "fire drug." Sulfur reacts first to the introduction of heat from a spark or flame. The yellow mineral ignites at a relatively low temperature, 261° centigrade. Its burning generates additional heat that ignites the charcoal and shatters the saltpeter, releasing its store of oxygen. This pure oxygen hurries the ignition of more fuel. Charcoal burns at a higher temperature than sulfur, emitting abundant thermal energy to accelerate the reaction further.

Any burning substance generates a quantity of gases that take up much more space than the fuel itself. When gunpowder ignites, these gases are created in an instant and the heat of the reaction makes them expand enormously. This rapid generation of hot gas produces all of the effects of "fire drug."

These chemical details, of course, were unknown at the time—the discovery of oxygen lay almost a thousand years in the future. Chinese alchemists devised theories from their own conception of universal dynamics, which viewed the world as a system of balanced dualities: Yin represented the passive, cool, female principle; Yang the active, hot, male. Interaction between the two generated the phenomena and transformations observed in the world.

"Saltpeter is the prince and sulfur is the minister," a sixteenth-century Chinese history text noted. "Their mutual dependence is what gives rise to their usefulness." Another document explained that salt-peter "is extremely negative or moonish (Yin). Sulfur is extremely positive and sunny (Yang)—that is, extroversive. When the two super-natural elements, Yin and Yang, meet each other in an exceedingly close space, the resulting explosion will stun every soul and shatter everything around it."

———

AS WE CAN see from the warning about the exploding elixir, practitioners initially viewed "fire drug" as a dangerous mistake. It didn't fit into the usual domestic functions of fire like cooking, lighting, and providing heat. It was an oddity to be avoided. For the moment alchemists lacked the tools to put the accidental discovery to work.

When ignited uncontained, gunpowder burns with a soft whomp and a burst of flame, leaving behind a cloud of dense white smoke—a magician's effect. This was perhaps its earliest application in China. Rather than heed warning of its dangers, alchemists may have spied an opportunity for parlor tricks that would mystify their audiences. The ceremonial and entertainment value of fire was highly esteemed in China, and "fire drug" offered a myriad of possibilities for the creative impresario. Pyrotechnics was the first field in which the magical new substance found a use.

Ignition translates gunpowder's stored chemical energy into the thermal energy of flame and the mechanical energy of compressed gases. Simple tools, containers of one sort or another, are needed to direct that energy and put it to work. It's likely that fireworks craftsmen designed the four basic forms of containment that have dictated all the uses of gunpowder from medieval China down to our own time.

First, enclose the powder in a container, sealed except for a fuse, and the gas generates enormous pressure, enough to blow the vessel apart. The tougher the container, the greater the energy that accumulates and the more violent the explosion. The sound of a firecracker results from the sudden tearing open of the paper under pressure and the expansion of the gas inside. A bomb is a container with a much harder shell.

Next, pack the powder into a tube with one open end—the combustion products fly out in a fiery spray. Pyrotechnic artists used the effect to produce spectacular fountains of fire. Bamboo offered convenient cylindrical containers for early gunpowder devices. These fire tubes were later the basis of some of the earliest gunpowder weapons.

When the gases rushed from those tubes, users felt a force pushing in the opposite direction. By accident or design, they began to put this force to work. The burning powder now drove the tube forward, turning it into the third basic tool, the rocket. The origin of the rocket is suggested by an incident in 1264 when the emperor Li Tsung hosted a feast in the Palace Hall in honor of his mother, Kung Sheng. During the festive fireworks show in the courtyard, the pyrotechnicians ignited "ground rats," small tubes that scurried around from the force of burning "fire drug." One skittered wildly across the floor and up the steps of the throne, frightening the Empress-Mother severely.

Insert the powder once more into a tube with one end open. On top of it, rest an object that almost completely fills the opening. When the powder burns, the expanding gases heave the object out the end. Under the right conditions, this tool efficiently converts the powder's chemical energy to mechanical force—the projectile bursts out at a high rate of speed. This is the fourth and most consequential of all the applications of "fire drug": the gun.

———

A DEEPLY ROOTED misconception in the West holds that the Chinese never used gunpowder for war, that they employed one of the most potent inventions in the history of mankind for idle entertainment and children's whizbangs. This received wisdom is categorically false. The notion of China's benign relationship with gunpowder sprang in part from Western prejudices about the Chinese character. Some viewed the Chinese as dilettantes who stumbled onto the secret of gunpowder but couldn't envision its potential. Others saw them as pacifist sages who wisely turned away from its destructive possibilities.

Confusing the matter was a long-standing conviction that the origins of gunpowder in China went back as far as 100 B.C. New inventions are often saddled with the names of existing technology—the Chinese word that designated exploding bamboo was given as well to gunpowder firecrackers, that of incendiary arrows to rockets. This practice can falsely suggest ancient origins. Incorrectly dating the discovery of gunpowder a thousand years too early gave the impression that centuries had passed before the Chinese developed gunpowder weaponry.

The fact was that Chinese military authorities quickly grew interested in "fire drug," and for a good reason. While the earliest development of gunpowder was taking place, China was facing a prolonged and unprecedented military threat.

The Sung emperors, who took power in A.D. 960, ruled over a dynasty that was culturally one of the most robust in all of China's long history, but militarily one of the most vulnerable. It was during the Sung era that "fire drug" became established as a major accomplishment of human technology.

Sung officials instituted a merit system based on rigorous testing to fill important government posts. They initiated land reform and equable taxation. Cities flourished—the original capital, Kaifeng, had three times the population of Rome in its glory. A later capital, Hangzhou, with more than a million inhabitants, startled Marco Polo—his native Venice, though one of the largest cities in Europe, boasted only 50,000 citizens. Movable type and the application of the compass to navigation were among the many technical advances of the Sung era. "The people of China are the most skillful of men in handicrafts," a Persian observer noted in 1115. "No other nation approaches them in this."

But these technical and cultural achievements unfolded against a background of military conflict. Tribes from the interior of Asia, mounted barbarians who had worried China's borders for centuries, were increasingly pressing into the empire. The reign of the Sung emperors was to end in the complete subjugation of their country by foreigners.

High culture, technical ingenuity, and pressing need combined to push Sung artisans to investigate "fire drug" with a sense of both fascination and patriotic fervor. In 1044 the Sung emperor Jen Tsung received from his adjunct a document entitled "Collection of the Most Important Military Techniques." This text contained two recipes for making "fire drug" that could be used in incendiary bombs to be thrown by siege engines. A third mixture was intended as fuel for poison smoke bombs. All were low in saltpeter, which meant that they relied on rapid combustion, not explosion, for their effects. They were the first practical formulas for gunpowder anywhere in the world.

Fire-starting was an obvious jumping-off point for gunpowder technology and these mixtures fit neatly into the long Chinese tradition of incendiary warfare. Arrows had been turned into incendiary weapons by attaching packets of burning pitch or other flammable material to the shafts. A container of gunpowder could create a much more vigorous and stubborn fire when it hit its target. In A.D. 969, during the early years of the Sung dynasty, a man named Yo I-fang received a gift of silk for developing a new kind of fire arrow, probably including gunpowder. By 1083 the Chinese were producing gunpowder fire arrows by the tens of thousands. To make the weapons, workers wrapped paper around gunpowder "in a lump like a pomegranate," attached it to the shaft of the arrow, and sealed it with pine resin. The archer lit a fuse projecting from the mass of powder before launching the arrow.

Perforated metal balls charged with gunpowder and equipped with hooks could cling to towers, ladders, and buildings, where they sprayed flame, igniting the structures. Catapults were used to hurl these incendiaries. The range was short—less than 150 feet—but the hissing, fire-spurting ball made a fearsome weapon.

When Emperor Jen reviewed the report cataloguing his unconventional weaponry, he may have grown concerned about the proliferation of "fire drug." A short time later, he forbade the export of salt-peter and sulfur. He soon banned all private trade in these strategic ingredients, turning gunpowder into a state monopoly.

———

SUNG TECHNICIANS were still investigating the military possibilities of gunpowder when, in 1126, the task took on a new urgency. In September of that year the Jurchens, a wild, semi-nomadic tribe from north of Korea, without written language or calendar, descended on the Chinese, laying siege to Kaifeng. The Sung warriors used their gunpowder arrows to defend the city, firing down from the walls at the enemy.

In the eighty years since they had recorded the first gunpowder formulas, Sung technicians had discovered a new use for the magical substance: the first explosive bombs. They fashioned these bundles of explosive from traditional materials like bamboo or paper wrapped with string. They probably borrowed the idea from the fabricators of fireworks—the bombs were in fact little more than large firecrackers. "Fire drug" was still too low in saltpeter to burst stronger casings. The purpose to which military men put the bombs was also similar to what firecrackers might achieve—to startle, frighten, and confuse the enemy. Sung soldiers tossed these "thunderclap bombs" from the ramparts of Kaifeng "hitting the lines of the enemy well and throwing them into great confusion." Containing three or four pounds of weak gunpowder each, the devices let out a resonant boom and an enormous billow of smoke. One observer noted the reaction: "Many fled, howling with fright."

But no fire arrows or frightening noisemakers could stay the onslaught of the Jurchens. In January of 1127, the Sung emperor surrendered his capital to the barbarians. The Jurchens carried the "son of heaven" north, dressed him in a servant's livery, reduced him to the status of a commoner, and forced him to live the rest of his life in exile. The sudden collapse of Sung power stunned the Chinese. The Jurchens swept into northern China, establishing the Chin dynasty. They readily adopted many Chinese customs—by 1150 they were making saltpeter "artificially" in manure heaps. The Sung Chinese crowned a new emperor, who retreated to the south with his retinue. He continued the dynasty only by relinquishing huge tracts of land and paying regular tribute to his Jurchen neighbors.

More trouble was brewing. Among the many bands of nomads who roamed the steppes of Asia was a ferocious but obscure tribe known as the Mongols. In 1206, after uniting diverse Mongol factions, a young man named Temujin was exalted by his followers and given the title Genghis Khan, "supreme commander." "Man's highest joy," he declared, "is in victory: to conquer one's enemies, to pursue them, to deprive them of their possessions, to make their beloved weep."

Genghis Khan would make many weep. Initially, the Mongols swept to the west, invading India, then Persia, the Caucasus, all of central Asia, and threatening even distant Europe. After Genghis died in 1227, his son Ogodei turned his attention to the east. The Jurchen, during the century they had ruled northern China, had assimilated into the more advanced culture of their subjects. The Chin dynasty now fell victim to an army of mounted horsemen even more fearsome than their own ancestors.

The Chin had learned the secrets of the Chinese gunpowder weapons and adopted them as their own. By 1231, when the Mongols attacked, powdermakers had devised a formula rich enough in saltpeter that its explosion could burst an iron casing. The Chin used this "Heaven-Shaking Thunder Crash Bomb" to defend Kaifeng against the invaders. The explosion of this blockbuster, a witness related, could be heard 33 miles away and scorched an area forty yards square. The shrapnel ripped through iron armor.

The Mongols tried to protect their sapping trenches by using cowhide covers, but the defenders lowered a thunder crash bomb on a chain. "The attacking soldiers were all blown to bits," an amazed chronicler noted, "not even a trace being left behind." The bombs were frightening indeed, but the point was no longer just to scare the enemy.

The desperate Chin defenders turned to another gunpowder device borrowed from their Chinese subjects, a spear or pike with a two-foot-long tube attached just below the point. This container, of bamboo or tightly wrapped paper, was packed full of "fire drug." A soldier lit the mixture and pointed it at the enemy, spewing flame and sparks as far as six feet. The eruption lasted five minutes. The lances were found to be "the best of all fire weapons." They were especially useful for defending city walls. "These thunder-crash bombs and flying-fire spears were the only two weapons that the Mongol soldiers were really afraid of," a chronicler recorded.

Not afraid enough. The Mongols defeated the Chin and ended their reign in China. The Sung Chinese, who ruled a large and wealthy kingdom to the south, must have known that their turn would soon come, but they failed to prepare for the onslaught. In 1257, a government official complained that Sung arsenals were woefully lacking in up-to-date weapons, especially iron bombs and fire arrows. The warnings were not heeded. Genghis's grandson, Kublai Khan, attacked China in 1274, quickly reducing the entire nation to an appendage of the vast Mongol empire.

———

LIKE THE JURCHENS before them, the Mongols were eager to make use of the proficiency of Chinese military technicians and craftsmen. They adopted "fire drug" weapons and encouraged their continued development. But they never viewed the radical technology as a replacement for the potent cavalry techniques by which they had conquered half the world. Gunpowder would remain a novel supplement to their traditional ways of making war.

On at least one occasion, they got a taste of the volatile and unpredictable nature of the new substance. In 1280, Mongol officials dismissed the Chinese workers who had been operating a gunpowder arsenal. The locals had fallen under suspicion "owing to their covetous and deceitful behavior" and had been replaced by Mongols. "But they understood nothing of the handling of materials," an observer noted. As a result, the arsenal exploded with much loss of life. It was an indication that though the Mongols were actively trying to adopt the new weapon, they still had much to learn.

Chinese engineers continued to develop gunpowder technology in spite of the danger. Their increasingly powerful bombs were as perilous to their users as to their makers. Heaving them from a catapult must have strained the nerves of the crew, who could not be sure of launching the bomb before it exploded. Timing was critical—the bomb had to have enough fuse to keep burning until it reached its target, but not so much that it could be extinguished or tossed back by the enemy.

During the thirteenth and fourteenth centuries, these bombs became common implements of war in China. Their names were fearsome in themselves: "Dropping from Heaven Bomb," "Match for Ten Thousand Enemies Bomb," "Bandit-Burning Vision-Confusing Magic Fire-Ball." One nasty device known as a "Bone-Burning and Bruising Fire Oil Magic Bomb" contained pellets of iron and shards of porcelain coated with a stew of tung oil, urine, sal ammoniac, feces, and scallion juice. The bomb was guaranteed to wound the skin and bones, blind enemies, even knock birds out of the sky. Both the names and the exotic ingredients indicate that an element of magic continued to be attached to many "fire drug" weapons.

Rockets also joined the arsenal, and were the most technically demanding gunpowder tools. With a firecracker or bomb, designers wanted the powder to burn as quickly as possible, generating an explosive burst of gases. A rocket required a regulated ignition that continued to provide thrust as the tube flew through the air. Technicians found that gunpowder, when packed very tightly into a tube, only burned from the surface. To provide thrust, they needed to leave a conical opening in the center of the dense powder—the enlarged surface area generated enough hot gases to lift the rocket skyward. Finally they had to invent a restricted opening that would channel those burning gases for more power. The result was a tube of gunpowder that could soar through the air and deliver an incendiary package on a distant target. Rockets probably began to be used in war in the middle of the thirteenth century.

———

AS REVOLUTIONARY as these rockets, bombs, and incendiary weapons were, Chinese technicians were soon working on an even more consequential tool: the gun. As they learned to make their gunpowder increasingly explosive, they saw new possibilities in the fire lance—the handheld flamethrowers that had long since become standard weapons. They found that the rush of gunpowder fire could carry with it bits of metal and crockery—the flying debris further injured and disconcerted enemy soldiers. Weapons-makers began to substitute metal tubes for bamboo. This allowed them to use a more powerful powder and to shoot a fiercer flame. A "Bandit-Striking Penetrating Lance" had a squat, yard-long iron barrel with a two-foot stock attached to the butt end.

Over the years, fire lances grew even larger. Armed with one, a soldier could blast at an enemy a six-foot tongue of flame accompanied by a nasty spray of metal splinters and broken porcelain. Eventually the devices became too cumbersome for one man to handle. Engineers attached them to wooden frames or to wheeled carriages. While not as versatile as the handheld version, these weapons, which have been called "erupters," would have presented an awful surprise to a band of marauders trying to scale a city wall or assault its gates. What's more, in using the lances to hurl missiles, the technicians were approaching a truly revolutionary breakthrough.

Gradually the projectiles became more crucial to the effectiveness of the weapon than the flames themselves. Some erupters shot bundles of arrows. One came with a magazine that dropped lead balls into the firing chamber to be spewed forth one after another. This weapon was said

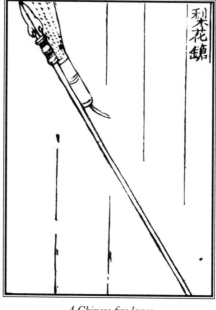

A Chinese fire lance

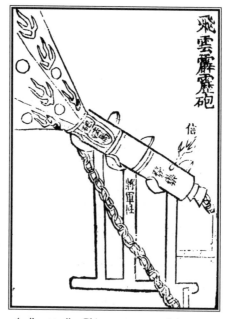

An "erupter," a Chinese prototype of the cannon

to be able to resist the assault of fifty soldiers. A "Nine-Arrow Heart-Piercing Magic-Poison Thunderous Fire Erupter" speaks for itself.

Weapons developers noted that the more closely the projectiles came to filling the diameter of the lance's barrel, the more forcefully they would fly from the mouth. As powdermakers devised formulas richer in saltpeter, the explosive powder threw the projectiles even farther. It also blew up the whole affair on occasion, pointing to the need for stronger tubes fashioned from thicker metal.

As the thirteenth century came to a close, with the Mongols still in control of China, this evolution of the fire lance resulted in an entirely new implement. Instead of using the flames of the incendiary powder directly, it harnessed "fire drug's" potential for throwing projectiles. This was the first gun.

For military technicians of the time, the new device was a logical variation on earlier weapons. From the broader perspective of mankind's endless quest for instruments of violence, it was one of the most monumental developments in history. In the true gun, the fire was secondary—the powder was a fuel, a source of concentrated energy. This energy was put to use indirectly, the real work being done by the projectile.

The earliest guns appeared in China in the late 1200s—the oldest extant hand cannon is tentatively dated 1288. The gun arrived in evolutionary steps, just as other Chinese gunpowder weapons had. The earliest were small, crude variations on the fire lance. A Chinese bronze hand cannon from 1332 is only a foot long and weighs eight pounds. Hundreds of larger artillery pieces have survived from the 1350s. They fired both stone and iron balls.

An account of a 1359 battle near Hangzhou, which had become a leading gunpowder-producing city, already takes the use of firearms for granted. The Ming Chinese were then challenging their Mongol rulers. Both sides were well-equipped with cannon, sometimes firing dozens of them in organized salvos. By 1412 the "Long-Range Awe-Inspiring Cannon," almost three feet long, used half a pound of powder to fire a two-inch ball at high velocity. Alternatively, it could shoot a bag of half-inch lead pellets.

The cannon took its place as a tool in the vast arsenal of the Chinese military. Engineers added improvements over the years. The weapon's effect, though, was far from revolutionary. It was another armament, another tool for using "fire drug." But the concept was about to travel to a land where it would create a far more profound effect, where it would not just serve as an instrument of war, but would upset an entire society and radically alter the course of history.

2

THUNDRING NOYSE

THE ENGLISH KING Edward III, a bold 34-year-old overreacher, traveled to the Continent in the summer of 1346 and mounted a campaign to enforce his claim to the throne of France. He landed on a Normandy beach and inflicted on the French countryside a form of violence known as "havoc"—murder, plunder, rape, war on the cheap. He burned towns within sight of Paris, alarming its citizens.

With his flowing blond hair and beard, his athletic physique, his devotion to pomp and pageantry, his love for the brutal spectacle of jousting, Edward embodied the age of chivalry. He hoped to make the mythical Camelot a reality, to preside over a Round Table as splendid as that of King Arthur. Yet his idealism and nostalgia never clouded his cold-eyed perception of war's realities. The king had brought with him a new technology that was all the rage across Europe: gunpowder.

Though he could hardly have imagined it at the time, this was the very weapon that would one day wipe away the customs of feudal knighthood that he held so dear.

War in the Middle Ages relied on man's muscles. A battle tested strength against strength—every warrior's pride was his ability to fight. Edge weapons like swords and spears concentrated muscular energy. Catapults and siege engines accumulated and stored human strength. The crux of battle was the melee, a free-for-all among men-at-arms. The sword, the extension of the arm, was the icon of war. Gunpowder would introduce a new dimension, one independent of human strength.

Gunpowder had been known in Europe for several decades by Edward's day, but had yet to find a solid place in the warrior's catalogue of arms. Embraced by brutal enthusiasts like Edward, it would, during the fourteenth century, begin to widen the scope of men's ability to inflict violence. It would force commanders to rethink military axioms that had stood for centuries and extend the reach and destructive capability of individual soldiers, setting off ripples that would be felt across Europe and around the world.

———

ONE THING IS certain: Europeans knew of gunpowder by the middle of the thirteenth century; how the technology reached the West has long been subject to dispute. In 1854 an historian asserted authoritatively that the Egyptians had used gunpowder and that Moses had known of it. A leading twentieth-century English expert on artillery listed seventeen arguments to show that "there is no trustworthy evidence to prove that the Chinese invented gunpowder." Instead, he maintained, they borrowed it from the West.

Three pieces of evidence have convinced scholars that Europe's knowledge of gunpowder originated in China. The first is precedence. The Chinese had some notion as early as the ninth century that saltpeter, sulfur, and carbon could burn with unprecedented vigor. By

1044 they had recorded formulas for gunpowder. The earliest reference to gunpowder in Europe is from 1267, the first formulas emerged around 1300, the first military use was noted in 1331. There is no evidence of gunpowder nor of any progress toward it in Europe until it had long been known in China.

The second persuasive indicator is gunpowder's lengthy evolution in China, including centuries of improvements in the refining of saltpeter, elixirs that flared up unexpectedly, and gunpowder formulations too weak to be explosive. This suggests that alchemists and military engineers proceeded slowly toward true gunpowder, then gradually strengthened it. No parallel development is evident in European records. Gunpowder appeared suddenly in Europe, and a little more than half a century later gunners were firing cannon at the walls of castles. This accelerated development indicates that Europeans borrowed a technology that had already been fully worked out in the East. No groping was necessary.

The third piece of evidence, minor but telling, is that early European gunpowder recipes included poisons like sal ammoniac and arsenic, the same ones that the Chinese used. These ingredients did nothing to improve the powder. Their presence in the formulas of both cultures is an unlikely coincidence and suggests that Europeans received the idea directly from China.

The route of this transfer and the exact date that gunpowder arrived in the West are not known and may never be known. Speculation has pointed in several directions. The Mongols were expanding across the entire Eurasian mainland in the thirteenth century. They swept through Persia and captured Baghdad in 1256. While they relied principally on their ferocious cavalry techniques, the Mongols brought Chinese engineers to western Asia and may have used gunpowder weapons against the Arabs. It's possible that they transferred to Europeans the secrets of the explosive.

Direct contacts between China and Europe, though still limited, were increasing at the same time that knowledge of gunpowder was spreading. Friars visited the Mongol court as early as the 1230s. Merchants and adventurers were also drawn East—Marco Polo did not return from the court of Kublai Khan until 1292, after gunpowder had already arrived, but other Italian merchants had journeyed to the Orient at midcentury.

Speculation has looked to an incendiary weapon known as Greek Fire as a possible precursor of gunpowder. This fiercely burning substance was invented in Byzantium around A.D. 675 by a Jewish architect and Syrian refugee named Kallinikos. Its formula was a closely guarded secret that remains a mystery to this day. Kallinikos most likely distilled petroleum into something like gasoline, which he thickened with resin to create a primitive form of napalm. It's possible that the incendiary contained saltpeter, which added to the intensity of its burning. In that case, gunpowder could trace a line of descent back to the Greek invention. Definitive evidence is lacking.

It's likely that the Arabs played some role in the transmission of gunpowder to the West. In the thirteenth century, the followers of Islam had established a cosmopolitan culture stretching from the Iberian peninsula to India with technical achievements surpassing anything in Christendom. Around 1240 the Arabs acquired knowledge of saltpeter ("Chinese snow") from the East, perhaps through India. They knew of gunpowder soon afterward. They also learned about fireworks ("Chinese flowers") and rockets ("Chinese arrows").

Arab warriors had acquired fire lances by 1280. Around that same year, a Syrian named Hasan al-Rammah wrote a book that, as he put it, "treats of machines of fire to be used for amusement or for useful purposes." He talked of rockets, fireworks, fire lances, and other incendiaries, using terms that suggested he derived his knowledge from Chinese sources. He gave instructions for the purification of saltpeter and recipes for making different types of gunpowder.

The oldest written recipes for gunpowder in Europe were recorded under the name Marcus Graecus or Mark the Greek. The attribution did not refer to an individual but to the editors who over two centuries compiled and emended a how-to manual entitled *Book of Fires for the Burning of Enemies*. The short work, written in Latin, very likely has Arabic roots—it may have been translated by scholars in Spain. "Fire flying in the air is made from saltpeter and sulfur and vine or willow charcoal," it states, giving proportions that amount to 69 percent saltpeter, which would have formed a relatively strong and explosive powder. The part of the manuscript dealing with the remarkable powder is a late addition, added between 1280 and 1300.

The introduction of gunpowder into Europe has traditionally been associated with two men. The first was Berthold Schwartz, known as Black Berthold or *der Schwartzer*, perhaps because of his complexion, perhaps to signify his interest in the dark arts. Some said he was a Dane or a Greek; most agreed he was a German; all were sure he was a monk. According to accounts from the fifteenth century, Berthold was an alchemist who heated sulfur and saltpeter in a pot until it exploded. He tried the same experiment using a closed metal vessel and it blew his laboratory apart. "The best approved authors agree that guns were invented in Germanie, by Berthold Swarte," an historian declared in 1605. The German city of Freiburg claimed Berthold as a native son and town fathers erected a statue of the great inventor.

The catch is that Berthold never existed—he was a legendary figure, like Robin Hood. The myth of his life served to bolster the Germans' claim of having invented the gun. The story also shielded Europeans from the fact that gunpowder, a critical force in their history, had emerged not from their own inventiveness but from the ingenuity of the Oriental mind. In fact, Berthold was an archetype, a stand-in for all the curious and ingenious experimenters willing to risk life and limb to explore and profit from the astounding new mixture.

The other seminal gunpowder figure is Roger Bacon, a real person and one of the most daring intellects of his age. Born around 1214, Bacon came from a wealthy English family and pursued an academic career at Oxford, then lectured at the University of Paris. In 1247, Bacon became intensely interested in the physical world and began to

A late sixteenth-century depiction of the mythical Berthold Schwartz.

examine natural phenomena in detail. He spent enormous sums on experiments in fields like optics and astronomy, attempting to build on the newly available theories of Aristotle. Often ornery and dogmatic, he crudely criticized other scholars. After becoming a Franciscan friar, he corresponded with Pope Clement IV, for whom he wrote three great works that were intended to sum up all human knowledge about the natural universe.

The story has long circulated that Bacon left behind a formula for gunpowder. It's said that he recognized the danger of the invention and so recorded the information only as an anagram, a code that remained unbroken for centuries. This is the stuff of legend and that's exactly what it turns out to be. The letter containing the alleged formula cannot be definitely attributed to Bacon, and the coded "formula" is open to any number of interpretations.

Bacon does hold the distinction of having set down the first written reference to gunpowder in Europe. It came in the works he prepared for the Pope around 1267—and which Clement died without reading. Bacon wrote of "a child's toy of sound and fire made in various parts of the world with powder of saltpetre, sulphur and charcoal of hazel-wood." The effect of the device was quite astounding to the medieval mind. "By the flash and combustion of fires," Bacon wrote, "and by the horror of the sounds, wonders can be wrought and at any distance that we wish—so that a man can hardly protect himself or endure it."

The potential danger of this new form of energy did not escape him. Since a tiny firecracker "can make such a noise that it seriously distresses the ears of men . . . if an instrument of large size were used, no one could stand the terror of the noise and flash. If the instrument were made of solid material, the violence of the explosion would be greater." It was a prophetic insight.

———

EIGHT DECADES later, in the French hinterland north of Paris, an English king was about to put to use the violent explosions that Bacon had foreseen. Edward's 1,200 men-at-arms and 8,000 longbowmen had been standing all day in the August heat. Many had encased themselves in garments of chain mail and plate steel. They gripped the smooth-worn handles of swords, knives, bludgeons. They waited in a field about twenty miles from the Channel coast—the nearby village of Crécy would give the coming battle its name.

The sky suddenly darkened, thunder rolled across the countryside. A crash startled a flock of crows and sent them wheeling over the slop-ing field just before a downpour. Anyone seeking a portent could have found it in the flight of black birds. Few persons alive in 1346 took such omens lightly. Death was in the air. Indeed, death was gathering a few hundred yards away: The French king Philip VI was massing thou-sands of trained killers astride warhorses at the far end of the field. A

Soldier firing a fifteenth-century hand cannon

line of Genoese crossbowmen were preparing weapons that could drive a bolt through a sheet of steel. The sun, slipping from behind a thunderhead, made the polished armor glitter and lit up the elaborate plumes and multicolored banners of the gathering warriors.

The foreboding of the English soldiers must have mixed with a sense of expectation as they waited for the debut of their unprecedented weaponry. The guns at Crécy were small cast bronze or iron tubes strapped to wooden frames. Perhaps they were "ribaudequins," collections of small guns arranged on a wheelbarrowlike cart that could fire a salvo. Edward did not assign them to soldiers. He needed specialists, men who understood this novel form of energy. Canny and daring, these gunners knew that the soldiers they accompanied looked on them with jaundiced eye. Hatred of innovation was a military reflex. Gunpowder, closely associated with the necromancy of alchemists, was dangerous, unreliable, and perhaps unlucky.

The gunners took hasty precautions at the approach of the thunderstorm to make sure the rain did not dampen their loaded weapons or their supply of "poudre." As they waited they experienced the gut-churning anxiety that precedes any battle, the more so because they could see that Edward's army was seriously outnumbered, menaced by the finest heavy cavalry in Europe. They knew that while a noble man-at-arms might be captured for ransom, a gunner or archer was sure to be hacked to death if the battle went against the English.

Though his imagination had been caught by the mystery of gunpowder, Edward remained a sober and skillful tactician. His principal means of contending with the French was an arm that had been thoroughly tested in battle: the longbow. Trained to the weapon from childhood, Edward's archers were men of tremendous strength and highly specialized skill. Their skeletons today show the signs of the overdeveloped muscles needed to draw a six-foot-long bow to more than 100 pounds of tension, hold it steady, let loose an arrow, repeat the process again and again, ten times a minute. They could hit targets up to 200 yards distant, their arrows slicing through chain mail and even light plate armor. Edward had forbidden all sport except archery in his kingdom; he could form up enough of these skilled bowmen to deliver a withering storm of projectiles.

To the aristocratic continental warrior, archers were cowards, attacking from a distance with their missiles rather than embracing the face-to-face combat that had been an emblem of military honor since the days of the Teutonic tribes. What was worse, they were commoners. The elites of Europe based their ascendency on military prowess, on their extravagantly expensive mounts, suits of armor, and fortified castles. To them, the notion of a plebeian slaying a gentleman was repugnant. It was an attitude they would apply with even more vehemence to those who wielded gunpowder weapons.

Horses could be heard neighing in the distance. The knights screamed insults back and forth. On the French side, the warriors were bragging about which English noblemen they planned to capture—their opponents' reputations were well known from the international tournaments. King Edward himself would have been the ultimate prize, followed by his sixteen-year-old son Edward, known as the Black Prince from the color of his jousting armor, who was now preparing to lead a unit of the army in his first great test.

As the sun drifted down behind the English line, Philip hesitated. The 53-year-old monarch was thrown nearly into apoplexy by the sight of this swaggering pretender. Yet Philip's army had barely formed, the Genoese were fatigued from forced marching. He had not had time to set down a clear plan of attack. Disorder reigned.

Philip's chevaliers would not be stayed. With evening approaching, fearful that the English would escape his clutches in the dark, he gave the order: Attack.

TO MOST OF the men gathered at Crécy, to all men only two generations earlier, the guns that Edward brought to the battle were inconceivable, elements of pure fantasy. It simply defied the imagination to propose that by no other action than the touch of a hot poker a man could blast a ball from the mouth of a tube and hurl it hundreds of yards at blinding speed. Anyone making such a claim would have been labeled a charlatan, a madman, or a sorcerer. Impossible. Nowhere in their writings had the ancients mentioned such a thing; nowhere in his fables and epics had man dreamed it.

Edward's new weapons were both the simplest tools ever invented and the most technically advanced products of the age. Like the proliferating array of guns that would appear throughout history, they consisted of little more than a tube sealed at one end, like a reed, or *canna* in Latin, "cannon." Edward's gunners placed their powder at the closed end, where a narrow hole let them introduce ignition. They deposited the projectile closer to the open end. At first they had fired an iron arrow or bolt, later they shot a lead or iron ball, or sometimes a stone chiseled into a sphere. The challenge of early guns came when the thrust of a red-hot rod down the touchhole set off the powder inside.

The explosion of the charge threw the same devastating pressure wave against the sides and rear of the tube as it did against the projectile itself, threatening to blow the gun to pieces. In essence, a gun was a bomb brought under control. The only material that could withstand the unheard-of stress and heat was metal. In the fourteenth century, metal remained a rare, expensive, and intractable material. Gunners were all too aware that a weakness at the breech (the solid base at the bottom of the bore), or an overload of powder, or a ball that jammed in the tube, could raise the pressure inside the gun beyond the bursting point, setting off an explosion that would spray the gunner himself with fire and jagged metal fragments. For the time being, the limits of metallurgy and scarcity of powder kept the guns small.

By the time Edward III mounted the throne, guns were becoming popular in diverse locations around Europe. The Italians, privy to trends in technology through their extensive trading contacts, adopted gunpowder during the years after 1300. The Signoria of Florence ordered city officials in February 1326 to obtain *canones de mettallo* and a supply of ammunition for the town's defense. That same year an English Chancery clerk named Walter of Milemete included the earliest known European picture of a gun in a fawning treatise he called "Of the Majesty, Wisdom and Prudence of Kings." The illumination, which is not discussed in the text, shows a vase-shaped container on a wooden trestle, a large arrow protruding, an armored man gingerly lighting the touchhole.

The new weapon became increasingly widespread during the 1330s. By the following decade most of the arsenals across Europe, from London to Rouen to Siena, listed some form of gunpowder weaponry. The guns were fired mostly in defense of town walls, but a reference from 1331 describes an attack mounted by two Germanic knights on Cividale, a town in the Friuli hills north of Trieste, using gunpowder weapons of some sort.

A French raiding party sacked and burned Southampton on the English coast in 1338, bringing with them a ribaudequin and forty-eight bolts. Since their supplies included only three pounds of gunpowder, they must have been more interested in showing off their new armament than in doing any serious damage.

Edward III joined this trend with enthusiasm. The London Guild-hall boasted half a dozen brass "gonnes," along with powder and lead shot in 1339. Two years before Crécy, the king lured Peter van Vul-laere, formerly master of ribaudequins at Bruges, to cross the Channel and supervise the preparation of English gunpowder weaponry. He hired several "artillers" and "gonners," to assist him. Among the supplies he sent to France to further his raids were 912 pounds of salt-peter and 846 pounds of sulfur for making gunpowder. Van Vullaere may have overseen the guns at Crécy.

A great deal was riding on Edward's innovative weaponry. A defeat could result in the capture or death of the monarch, radically altering the fortunes of the nation. With the Black Prince also on the field, the French could erase an entire dynasty with a stroke. Thousands of Genoese mercenaries were now advancing up the slope. Their role was to come within striking distance, a hundred yards or so, and let loose their bolts into the English ranks to soften them up for the impending charge of the French knights. The fate of a kingdom rested on the outcome.

————

WAR IS A psychological drama as well as a physical contest. The goal of battle is to shock, to shatter the cohesion of groups, to bring individual soldiers up short, to instill fear, sow confusion, undermine morale, sap will. Violence is one means of doing so. Intimidation—through a display of power or a loud noise—is another.

Sound was always an accessory of war—drums, trumpets, bag-pipes. Shouting was universal. The crossbowmen at Crécy raised three resounding shouts as they advanced within shooting range. The English met each of these bellows with silence. They simply waited.

We don't know exactly when Edward chose to fire his guns. One account relates that the English "struck terror into the French Army with five or six pieces of cannon, it being the first time they had seen such thundering machines." Another states that they fired "to frighten the Genoese." A third says they "shot forth iron bullets by means of fire. They made a sound like thunder." The thunder that met the at-tackers made their bellicose shouts seem feeble by comparison.

Astound. Astonish. Stun. Detonate. All of these words derive from a root meaning thunder. On firing, Edward's guns shot out great tongues of flame followed by roiling clouds of white smoke, an im-pressive and unique sight to the French knights and their allies. More astounding, more stunning, was the powerful sound of the detona-tions. If, as one chronicler noted, it scared the horses, it most certainly startled the men as well. It was thunder brought to earth, sound hurled forth as a weapon. Like the crash of thunder from a nearby lightning strike, cannon fire from close range is heard not just with the ears but with the gut, the bones, the nerves. It is feeling more than sound, a sudden expansion of air that delivers a physical blow.

Writers who described early guns almost invariably compared their sound to that of thunder. "As Nature hath long time had her Thunder and Lightning so Art hath now hers," one observer noted. Shake-speare called them "mortal engines, whose rude throats th'immortal Joves' dread clamours counterfeit." Early names for guns referred to their booming sound. In Italian they were *schioppi*, or "thunderers." The Dutch had their *donrebusse*, "thunder gun," in the 1350s. In English this became the blunderbuss.

The term "gun" had a different origin. The word most likely derived from the Norse woman's name "Gunnildr," familiarly shortened to "Gunna." "Gonne" first shows up in a 1339 document written in Latin. Geoffrey Chaucer, who served in Edward III's royal administration, initi-ated the vernacular use of the word in 1384:

As swifte as pelet out of gonne
   When fire is in the poudre ronne.

———

EARLY GUNPOWDER was weak, and the light guns were unreliable and inefficient—they could shoot only small bits of metal and their accuracy was atrocious. Reloading was cumbersome and time-consuming. All of these factors made the effect of their missiles almost inconsequential. The guns at Crécy struck few men from their horses.

So what led rulers like Edward III to invest scarce resources in the manufacture of guns and the grinding of powder? What induced them to pursue the new technology with such unquenchable enthusiasm? Surely guns possessed a mystique that went far beyond their military effectiveness.

Their diabolical associations were potent. A reputation as a friend of the devil was a valuable asset on the battlefield. The men at Crécy believed wholeheartedly in all the lurid imagery of Christian metaphysics. Hell was a real place, choked by burning brimstone. Demons strode the earth. The guns' sulfurous exhalations, hideous roars, and stark flashes of light were all trademarks of Satan.

In the secular realm, gunpowder weapons were power made manifest. The man who could field guns, like he who rode the most expensive horse, was a man to be reckoned with. The accouterments of war carried their own prestige, and the medieval mind was deeply taken by regalia. Gunpowder added to the dramatic elements of battle—first as a minor stage effect, later as a dominant theme.

So Edward, in spite of debts that had already bankrupted him, purchased his guns and his powder. Battles, of course, are ultimately won not by show but by violence. Edward's use of well-positioned archers and armored men fighting afoot from a strong defensive position proved insightful. If the guns gave the Genoese a serious start, the hail of arrows from the English longbows sliced into their ranks with devastating effect. The whizzing arrows cut into horseflesh and occasionally found a gap in a knight's armor. The disciplined English line withstood repeated attacks. In charge after charge, the standard-bearers of French chivalry were butchered. Philip barely escaped the field with his life. As darkness descended, his Bohemian ally, King John the Blind, was led into the fray by companions—he wanted to die fighting. He did.

———

THE GUNPOWDER that Edward brought to the field at Crécy was a precious and poorly understood commodity. The men who assembled its materials were the brothers of bakers and brewers: They devised their methods through intuition and knew that minor variations in procedures could significantly alter the outcome. The craft attracted adherents across Europe: alchemists, blacksmiths, enterprising peasants, men fascinated by the unknown or intrigued by the commercial potential, daredevils, visionaries, madmen. Some found in the profession not fortune but disfiguring burns and death—the grinding of gunpowder was ever a perilous craft.

Sulfur, familiar from Biblical times, was the simplest ingredient, easily purified and ground to a fine powder. Charcoal, long used for cooking and metal work, was also easy to obtain. The species of the tree from which it was made mattered most. Charcoal for gunpowder needed both a delicate structure that made it easy to pulverize and a minimum of ash content. Willow was a common source. Alder, chinaberry, and hazelwood also worked, as did grape vines. Old linen sheets heated in a closed vessel were used. In China, adding charred grasshoppers was believed to give powder liveliness.

In Europe, the difficulty of collecting saltpeter long remained a bottleneck in gunpowder production. The continent lacked the hot climate that encouraged rapid decomposition and the extended dry period that allowed nitrates to leach to the surface. Gunpowder fabricators had to seek out saltpeter wherever they could find it.

Medieval Europe was much more rank than our sterilized twenty-first century. Peasants—the vast majority of the population—shared dirt-floor hovels with farm animals. Food scraps and dog shit were ground into the reeds that served as carpeting. Night soil and manure were the only fertilizers, open sewers the norm in towns. It was from this stinking fundament of the human environment that gunpowder makers derived their most precious ingredient.

Observers saw *sal petrae,* "salt of stones," forming in white crusts on stone walls. An early monk described it as a "wonder salt" with an infernal spirit concealed in icelike crystals. A writer in 1556 said that saltpeter could be "made from a dry, slightly fatty earth, which, if it be retained for a while in the mouth, has an acrid and salty taste." Saltpeter had long been used as a preservative to help meat retain its redness. Physicians prescribed it for ailments like asthma and arthritis. In fact, it can be toxic in large quantities, causing anemia, headache, and kidney damage. It was at times touted as an aphrodisiac, though persistent rumors have also insisted that the overseers of army barracks and boys' schools snuck it into the food to quell the carnal appetites of their charges.

The salt formed on the walls and floors of privies and stables, and in "Vaults, Tombs, or desolate caves, where rain can not come in." But the natural supply was meager. Every kingdom was hard pressed to produce enough of the crucial substance. Powdermen scoured the land looking for old manure piles and cesspools, middens, and pissoirs. Collectors armed with royal warrants scraped saltpeter from barnyards and dovecotes. Their intrusions annoyed residents who, having seen their yards dug up and structures demolished, were required to provide lodging for the petermen and fuel for boiling down the smelly liquid leached from the ordure.

In 1670 a gentleman named Henry Stubbes mentioned a cave in the Apennine "in which Millions of Owles did lodge themselves, their dung had been accumulating there for many centuries of years." Mining the guano for its saltpeter produced an "inestimable summe of money." Around the same time, the bodies of soldiers hastily buried in caves after a battle near Moscow proved a rich source of saltpeter, which was fashioned into new powder to facilitate the deaths of other men, a macabre kind of recycling.

Having observed where they could find saltpeter in nature, craftsmen of the late fourteenth century began to create the same conditions artificially. These attempts to hasten the decay of organic materials and prevent the runoff of the nitrate salts developed into saltpeter planta-

A 1598 depiction of a saltpeter "plantation" and refinery.

tions, which were an elaboration on the simple compost heap. The first record of one is in 1388 Frankfurt. By the following decade, artificially produced saltpeter was fueling a more abundant supply of gunpowder.

The process was not difficult—anyone with a covered pit or cellar and a supply of manure could get into the business. A saltpeter recipe from 1561 suggests mixing human feces, urine, "namely of those persons whiche drink either wyne or strong bear," dung "of horses which be fed on ootes," and lime obtained from old mortar or plaster. The knee-deep pile was to be kept sheltered from rain and turned regularly for a year. Saltpeter would then emerge "like snow." The prescription for the piss of boozers was not fanciful—the metabolization of alcohol produces urine rich in ammonium, a food that nitrate microbes thrive on.

Gunpowder makers had to process a hundred pounds of scrapings to yield a half pound of good saltpeter. Workers leached water through the foul mass to dissolve the nitrates, then crystallized them out of the resulting solution. Here they ran into a problem. The best form of saltpeter for making gunpowder was potassium nitrate, but most of the nitrate salts produced in nature were those of calcium. Calcium nitrate served well for fashioning the explosive, but it had a quality that created difficulties later—it absorbed water from the air, eventually rendering powder damp and unusable. The gunpowder that European craftsmen made in the fourteenth and early fifteenth centuries included a large proportion of calcium nitrate and the problem of spoilage by damp was widespread.

The plantation production of saltpeter, known as "petering," became a cottage industry in some countries of Europe, a part-time occupation for anyone who could tolerate the stench. Plantations made larger quantities of powder available and played a role in the spreading use of gunpowder weapons in the fifteenth century.

The powdermaker, once he had assembled the three ingredients, had to grind them together in a mortar. The proportions were important, but recipes of the time already approximated what are now known to be the ideal percentages: 75 percent saltpeter, 15 percent charcoal, 10 percent sulfur. "There is a certaine proportion of Perfection, of these three Components," an early observer noted, "And that in such sort, as if you adde more or lesse Petre, the Violence shall abate."

Intimately mixing the ingredients could take a day or more of relentless pounding. In the process, these three harmless, naturally occurring substances combined physically, not chemically—each retained its nature and could be separated out. Yet the common ingredients took on a new life, an edgy and esoteric relationship with fire, an ability to explode with the greatest violence.

———

EDWARD III paced the battlefield at Crécy the morning after the fight, surveying the carnage. A French herald accompanied him to help identify the dead. King John of Bohemia. The Count of Lorraine. The Count of Flanders. Barons, earls, noblemen of the highest rank,

Fourteenth-century craftsmen grinding gunpowder

knights by the hundreds. Certainly the fight had been, as one chronicler noted, "very perilous, murderous, without pity, cruel and very horrible." The result was a shock to the French—even the fiercest pitched battles in medieval times had rarely seen such slaughter.

Gunpowder had played mainly a psychological role in the fight, frightening enemy soldiers and horses, boosting English morale, adding to the confusion in the French ranks. It had yet to establish a dominant position in war, but Edward and the rulers of his generation had glimpsed the possibilities in this novel and unique form of concentrated energy.

The war Edward had started was far from over. It would go on beyond his lifetime, beyond the lifetimes of his children's children. Northwestern Europe, during the protracted struggle of the Hundred Years' War, would be a sorry place, wracked by spasms of violence even as the Black Death robbed the continent of 40 percent of its population. During the desperate competition over who would rule France, a steady parade of kings would probe the possibilities of gunpowder.

3

THE MOST PERNICIOUS ARTS

FROM CRÉCY, Edward III marched his army north to the port of Calais. Its citizens withdrew behind their stout walls, closed their gates, and steeled themselves for a prolonged test of wills with the English foe. Thus began one of the most common forms of medieval warfare: the siege.

During the Middle Ages a castle or town wall could thwart almost any attacker. For the besieger, prolonged encirclement and starvation of the trapped populace were the keys to victory, but laying a siege was a ruinously expensive undertaking. The encircling army needed to maintain constant vigilance against a relieving force or a raiding party sallying from the city itself. Wells and strategic supplies of food allowed a city or castle to withstand isolation for many months. With no means to penetrate stone walls, the besiegers often had to give up and go home.

Castles had been going up all over Europe since the eleventh century. The Normans had secured their possession of England by erecting more than 900 castles around the island. Relatively easy to build and resistant to most siege techniques, castles encouraged medieval nobles and strongmen to assert their independence, resulting in the centuries of localized war that plagued feudal Europe.

From Roman times, military commanders had attacked walls with destructive "engines." They used catapults, which heaved stones; *ballista,* which resembled oversized crossbows; and other contraptions to attack enemy fortifications. They now had in their possession a new, chemical-driven engine that could throw its ammunition at a higher velocity than any mechanical device. It could fire on a level trajectory, point blank, rather than lob stones the way catapults did. Its ammunition could slam directly into a wall or gate, perhaps breaking it.

The castle's strength resulted from the height and hardness of its walls. Attackers could not bore through stone, and defenders had an impregnable vantage point for showering enemies with missiles. Yet these very strengths made the walls a perfect target for guns. Stone masonry, hit with enough force, cracked. Once the structure was broken, its height became a disadvantage—the wall's high center of gravity threatened to collapse the entire structure.

Edward's gunners lined up twenty "gunnes gret and other gret ordinance" to batter the walls of Calais. The blasts and the shock of the stone balls crashing into the walls disheartened the residents inside. As autumn wore into winter, the repeated booms jangled their nerves.

But the guns at Calais, like those at Crécy, ultimately had little effect. Edward's imagination outpaced his technology. His gunpowder was too weak, the balls that issued from his pieces too small and too slow to smash the bastions of the French city. With the guns unable to break the stalemate, the siege progressed in the usual manner. Inside the city, supplies dwindled and the residents resorted to eating rats. In August 1347, after King Philip shied away from a relief attempt, the town's burghers sued for peace. They marched out with rope halters around their necks, offering themselves to be hanged to redeem their fellow citizens. Edward spared the townspeople but banished all citizens of rank and replaced them with English merchants and their families, solidifying control of the city.

———

THOUGH GUNPOWDER had proven ineffective at Calais, gunners had glimpsed their mission. The gun was to be the ultimate siege engine, the definitive wall-breaker. Gunners, powdermakers, and metallurgists went to work. They had cast the small guns of the early 1300s from either bronze or iron. But casting metal was a highly skilled, technically challenging, and enormously expensive craft. Large cast guns would arrive only slowly. In the meantime, wrought iron offered the greatest promise. Blacksmiths had long experience working with the tough, malleable metal. All they needed was a technique.

They borrowed it from the cooper's staves and hoops. They formed long wrought iron rods, heated them, aligned them along a wooden pole that served as a core, and hammered them together. When they had completed the circle they filled any cracks with lead and slipped white-hot iron rings over these staves. The rings shrank as they cooled, forming a durable tube, a "barrel." This type of gun looked like a section of water main or sewer pipe—it lacked the elegant tapered shape that we associate with the classic cannon. To make a bigger gun, the fabricator simply started with a thicker core and added more staves and larger rings. This hooped cannon was a major technical breakthrough. Blacksmiths, using the age-old methods of the forge, could now fashion pieces of ordnance whose size was virtually unlimited.

Guns quickly grew to astounding proportions. These superguns, known as "bombards," soon spread across Europe. In 1388 Nuremberg gunners christened a monster gun with the woman's name "Chriemhilde." Her three-ton barrel fired a stone ball carved from marble that weighed more than 500 pounds

Moving an object of such mass presented immense difficulties. First the gun crew had to hoist it onto a specially reinforced wagon using

A bombard in its wooden cradle

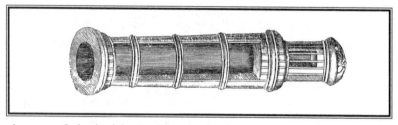

A cutaway of a bombard showing the powder chamber

ramps and pulleys. Teamsters then coordinated the slow progress of the piece along the roads. Marred by ruts and potholes in dry weather, the routes became sloughs of mud in wet. Bridges creaked under the load. One reason the dukedom of Burgundy led the continent in the adoption of cannon was because its system of rivers and canals facilitated transport.

A critical component of every gun was the powder chamber, a container with a thick wall and reduced bore that fit into the breech or rear end of the gun. Some weapons included a chamber as long as the barrel itself, others used a separate vessel shaped like a beer tankard. The gunner loaded this container with a charge of powder, then hammered a wooden bung into its open end. To fire his cannon he fastened the chamber to the rear of the gun barrel, screwing it with levers that fit into capstan holes, or holding it securely with a wooden or metal wedge that pressed against the back of the cradle. The touchhole or vent allowed him to insert the hot poker that fired the gun. The powder took fire, the bung let go, the hot expanding gases pushed the gunstone before them as they rushed up the barrel.

The bombard was the first really effective European tool for using gunpowder. The population of a besieged city first had to endure the earth-shattering explosion of the gun. Then came the thud of the ball, the trembling of their stone walls. In 1382 an army besieging Oudenaarde "made a marveylous great bombarde shotyng stone of marveylous weyght and when this bombarde shot it made suche a noyse in the goynge as though all the dyvels of hell had been in the way." The new weapons chipped away at the defenders' nerves even as they weakened and cracked stalwart fortifications. No kingdom wanted to be left behind in acquiring a weapon of such destructive potential.

Meanwhile, gun makers continued to push the limits of metal. Philip the Good, Duke of Burgundy, ordered artillery merchant Jean Cambier to assemble a gun with a barrel 15 feet long. This monster weighed almost eight tons. Gunners packed more than a hundred pounds of powder into the chamber to fire a 500-pound stone. The feverish gun makers of Flanders turned out a bombard called Dulle Griet or Mad Margaret. Fashioned around 1450, she tipped the scales at 18 tons. Her 16-foot barrel blasted stones two feet across.

Mons Meg, another of Cambier's forgings, had a bore diameter of more than 19 inches. The barrel was fashioned from iron bars two-and-a-half inches thick and wrapped with hoops made from three-and-a-half inches of solid iron. Philip the Good sent this piece to his Scottish allies in 1457. Its recipient, James II "of the fiery face," was an artillery enthusiast who used cannon in his festering hostilities with the English. His zeal proved unlucky. In 1460, the 30-year-old monarch, "more curious nor became the Majestie of ane kinge," as one observer

Cannon and fire arrows in action

noted, stood too close to a cannon during the siege of Roxburgh and was killed when it blew up beside him, a grievously common occurrence in the early days of artillery.

——

BY THE FIRST decade of the fifteenth century, guns had evolved to such a state that no commander would neglect to bring them along on a campaign. Their novelty had long since worn off. The Italian poet Petrarch had written as early as the 1350s: "These instruments which discharge balls of metal with most tremendous noise and flashes of fire ... were a few years ago very rare and were viewed with greatest astonishment and admiration, but now they are become as common and familiar as any other kind of arms. So quick and ingenious are the minds of men, in learning the most pernicious arts."

In 1415 the English king Henry V continued the war begun by his grandfather Edward III. Henry came to Normandy with the notion of capturing Harfleur, which controlled the Seine estuary. He set to battering the walls with ten guns. Three of them were bombards substantial enough to bear the names London, Messenger, and The King's Daughter. They terrified the inhabitants with their noise, smoke, and with the sickening crash of 400-pound stones into the solid walls.

"And the nimble gunner/With linstock now the devilish cannon touches" was how Shakespeare described the scene for audiences almost two centuries later, "And down goes all before them." Henry's gunners fired into the gates of an important outwork, smashing them and chasing the defenders away. There was no need for the English to rush through the breach. The roar of the guns had shattered their opponents' morale. Having witnessed violent destruction unlike any they had known, the townspeople agreed to Henry's terms. The siege had lasted only six weeks.

A decade later, the French were in desperate straits. The sickly and irresolute Charles VII had inherited an impossible situation. The English, riding the conquests of Henry V, had established themselves as

Mons Meg, a fifteenth-century bombard

masters of much of France, with the Duke of Bedford ruling from Paris. The Burgundians had thrown their power behind the English. As dauphin, Charles had retreated below the Loire River with his supporters, hoping to regroup. In 1428, Bedford decided to extend England's dominion south. He attacked and laid siege to Orleans. Conquest of the city would bring central France under English control. Only a miracle could reverse Charles's fortunes.

The miracle happened, an historical event as magical as any fairy tale. A 17-year-old peasant girl appeared. Beset by celestial "voices," she signed her name Jehanne and was known in her time as La Pucelle, the Maiden. She has lived in history as Joan of Arc. Drawing on a single-minded inspiration and an astounding knack for military tactics, she rallied the dispirited French troops, broke the siege, inspired the army to victory over the English, and arranged for Charles to be crowned king in the ancient capital of Rheims, cementing his legitimacy, stirring the nation, and reversing generations of defeat. Her short, fabulous career proved again that inspiration, morale, and superstition could play decisive roles in military struggles.

But Joan's practical contributions to the fighting were equally significant. More than one witness at her posthumous rehabilitation trial spoke of her talent for placing and using gunpowder weaponry. "She acted so wisely and clearly in waging war," said one supporter, "as if she was a captain who had the experience of twenty or thirty years; and especially in the setting up of the artillery, for in that she held herself magnificently."

Legends cling to Joan like the butterflies that allegedly hovered around her standard, but the evidence of her military inventiveness and her special skill at using gunpowder is strong. Her abilities with the revolutionary chemical energy are suggestive of two facets of gunpowder in the second quarter of the 1400s. First, the practicalities of gunpowder's use were in the hands of the artisan class, commoners. The danger and diabolical trappings connected with the mixture probably attracted an odd collection of visionaries and pioneers to the craft. It certainly required practitioners who were comfortable with risk and willing to get dirty in their work. Like the blacksmiths of the day, they were a peripatetic lot, ready to travel to where the work was, eager to sell their expertise to the highest bidder. Joan, a peasant herself, was free to mix with these gunners, to discuss with them on an equal footing the fine points of powder and its use. Such access to information would have been valuable in helping her to gain a sense of the technology's capabilities.

The second facet of gunpowder was its novelty, and her youth also proved an advantage in mastering a form of energy that was beginning to upset long-held military precepts, and whose use was rapidly evolving. Military commanders versed in classical theories, for whom gunpowder weaponry was an awkward intrusion, struggled to incorporate the guns into their strategic thinking. Joan, lacking preconceptions, viewed artillery with fresh eyes and readily developed an intuition about its use. Her facility has its distant echoes in the youth of the present day who quickly grasp the technology and possibilities of computers while their parents struggle to catch up.

Joan's mission was brought up short when she faced the Burgundian forces at Compiègne. Duke Philip the Good, having captured the Maid in a skirmish, handed her over to the English. Religious sensibilities of the day required that her visions be officially branded demonic—such was the result of her subsequent show trial. The penalty for heresy was death. Charles, whom she had helped gain the throne, did not act vigorously to secure her release. In May of 1431, not yet 20 years old, Joan was burned alive in the marketplace of Rouen.

Firmly seated on the throne, Charles VII began to act like a king, proving an adept administrator. Impressed by the formidable ordnance of the Burgundians, who deserted the English and joined the French cause in 1435, Charles hired two brothers, Jean and Gaspard Bureau. They created the world's first full-fledged artillery organization. French gunners, who up to now had invariably been independent contractors, were grouped into structured units, even given distinctive uniforms to wear in parades. Guns were no longer a stage prop but were becoming an indispensable part of war. Gaspard Bureau, *maitre de l'artillerie*, oversaw thirty cannoneers, a keeper of artillery, a master carter, and other professional technicians.

Marching through Normandy, the French army laid siege to Rouen in October 1449. The guns opened up. The city surrendered after only three days of bombardment. Cannon battered Harfleur into submission two months later and returned it to French hands for the first time since the conquest of Henry V. It had taken Henry six weeks to reduce the fortress—it took Charles only seventeen days.

Charles carried the war into the regions around Bordeaux, which the English had held since the twelfth-century reign of Henry II. Guns proved decisive again and again. Some towns surrendered immediately rather than face bombardment. The last gasp of English resistance came when 80-year-old Sir John Talbot led a force to reestablish English sovereignty in the area. Jean Bureau mounted a siege of the English-held town of Castillon. On July 17, 1453, Talbot led a force to relieve the town. Bureau turned his siege guns around and blasted the approaching English ranks. According to a French witness, "The artillery . . . caused grievous harm to the English, for each shot knocked five or six men down, killing them all." Talbot himself was blown to pieces. The French entered the city "through breaches made by the artillery." Castillon surrendered. The dynastic war that had brought grief to Europe for more than a century was over.

———

IN THE SAME year that the bombards of the Hundred Years' War fell silent, gunpowder was sending ominous reverberations from the other end of the continent. Constantinople, the Queen of Cities, had for eleven hundred years stood guard at the divide between Europe and Asia. But the rising Ottoman empire had long threatened to overwhelm the ancient capital. The danger came to a head in 1451 when the Turkish Sultan Mehmed II broke relations with Emperor Constantine XI, last heir to the eastern Roman Empire. Mehmed had already established his dominion beyond the Bosporus, winning control of most of Greece and a chunk of the Balkans. He was determined to turn the illustrious and strategic metropolis into the jewel of his empire.

Mehmed was patient but violently moody, famous for his cruelty as well as for the delicacy of his poetry. His face, with its arched brows, long hooked nose, and sensual lips, was said to "remind men of a parrot about to eat ripe cherries." He was only twenty when he set out to take Constantinople. The project came to obsess him. Unable to sleep, he spent his feverish nights plotting a way to capture the prize.

The task was a daunting one. Constantinople had survived twenty sieges over the centuries. The city—really a compact peninsula that included farms as well as urban settlements—boasted the most impressive fortifications in the world and was widely considered impregnable. Its double walls dated back to the fifth century. The inner wall was 40 feet high. In front was a stretch of cleared ground, then a 25-foot wall fronted by a 15-foot-deep ditch. Properly manned, the walls could withstand virtually any onslaught.

Little is known about the gunner named Urban who played a key role in the fate of the city. He reportedly came from Hungary, a country whose rich ore resources had helped to give it a lead in metal work. Judging by results, Urban must have been among the most skillful metallurgists and gunpowder technicians of his time. He offered his services to the Byzantine emperor. The salary the Greeks promised was niggardly, the supply of metal available for fashioning cannon meager. Apparently feeling no loyalty to the Christian cause, Urban sounded the Moslems as to what they might pay. Mehmed listened carefully to Urban's proposal. He asked him if he could fashion a gun to break down the walls of Constantinople. Eager to close a sale, Urban asserted that he could. Mehmed rewarded him with riches beyond what he dared ask. Urban set to work.

Versed in the techniques still being developed at the time, Urban knew how to cast on a large scale. Solid castings of copper alloys were required to withstand the pressure needed to drive the projectiles that could break the walls of Constantinople. Urban spent three months crafting an oversized cannon. He melted copper and added a small percentage of tin for hardness. Some of his material came from ore, much of it from melting smaller guns—Mehmed was willing to bet part of his valuable artillery train on Urban's ability to produce a supergun.

That Urban could achieve a precision casting on such a scale using a makeshift foundry is one of history's most impressive engineering feats. He made his giant bombard in two parts that screwed together. The rear section was a chamber thick enough to withstand the explosion of a massive quantity of powder; the barrel's bore was large enough to accept an enormous stone projectile.

Mehmed was delighted with the new gun. He mounted it on the fortress he had provocatively built overlooking the Bosporus. He issued orders that no ship should pass without his permission. A Venetian merchantman tried to defy the edict—Urban's gunners fired the bombard and sank the vessel with one lucky shot. Mehmed had the crew beheaded, the captain impaled. The Venetians and Genoese were thoroughly alarmed—their lucrative trade with Black Sea ports hung in the balance. The Genoese sent a fleet and 700 soldiers to support Constantine. This was merely a gesture. The idea of losing the city was unimaginable, yet the fractious Christian principalities of Europe could not muster a serious effort to confront the danger.

Mehmed, impressed by the skill of the gunners, ordered Urban to outdo himself. In January of 1453, the Hungarian cast another gun. This one was truly monstrous in its proportions: the barrel 26 feet long and capable of throwing a stone ball weighing more than half a ton. Fifty yoke of oxen could barely move the giant piece. Seven hundred men were assigned to the crew that would operate it. Urban gave a day's notice before his first test firing so civilians would not panic. Persons many leagues away heard the thunder of its detonation. The stone flew a mile and buried itself six feet into the earth. By April the massive gun was installed in front of the Sultan's tent along with numerous smaller pieces, all aimed at the walls of the city.

Mehmed led a force of an estimated 80,000 men, including 20,000 bashi-bazouk irregulars. Twelve thousand elite Janissaries formed the core of his army. The Sultan recruited these troops from Christian families as youths, converted them to Islam, raised them as skilled and fanatical fighters, and assigned them to his personal service. Inside the city, Constantine could count on only 6,000 native soldiers—the final remnant of an army whose roots stretched back to imperial Rome—and 3,000 foreigners, including some idealistic Spanish knights who had come to act out their dreams of chivalry defending the faith.

Mehmed gave the Byzantines a chance to surrender and be spared. They refused. On April 12 the bombardment began. The huge balls crashed into the walls, shaking them, crushing stone against stone. The gunners worked tirelessly. Gunpowder was delivered by the ton. Loading the huge cannon properly took hours. The gunners could fire the largest piece only seven times a day. Some of the balls thrown against the city were carved from marble taken from the temples of ancient Greece.

With time, the relentless pounding began to break down the outer wall. The defenders tried hanging plank barricades or bales of cotton over the ramparts to absorb the force of the balls. At night, when the barrage ceased, they frantically shoveled dirt and rubble into the damaged areas and erected wooden stockades to block the worst gaps.

On the night of May 28 the Byzantines noticed lights burning brightly in the Turkish camps. The Turks were dragging the guns even closer to the walls. Others were throwing tree branches, earth, fascines of bound twigs into the defensive ditch. At 1:30 in the morning, drums and cymbals sounded. The guns opened up, pounding the walls. For four hours a massive cannonade threw its echoes across the dark Bosporus. Alarm bells clamored throughout the city.

At dawn, masses of Janissaries approached the ditch. A giant named Hassan climbed to the stockade and fought fiercely with the Byzantines. He forced them back before being cut down. Some Turks sneaked into an unguarded side gate and planted their flags on the walls. Confusion caught hold of the exhausted Byzantines. Suddenly, the Janissaries were over the first wall. Maintaining their ranks, they pressed the defenders to the main wall beyond. Unable to retreat further, the soldiers were annihilated. Their fellows atop the walls gave in to the mounting panic, running for their homes beyond, hoping to protect their families. Scaling ladders went up unopposed. The Turks entered the city, threw open the Military Gate of St. Romulus. Dawn was breaking. Amid a pungent haze of gunpowder smoke, Constantinople was taken.

Mehmed had promised his troops three days of plunder—it was the traditional fate of any besieged town captured by storm. Chaos swept through one of the greatest cities in the world as looters raced ahead of the regular troops to grab booty and prisoners.

Christian witnesses left behind lurid accounts of the pillage, of the streets running with blood and severed heads bobbing at the shoreline. Terrified residents gathered in the ornate cathedral, Hagia Sophia, and prayed for a miracle. Their answer was the thud of battering rams against the doors.

Yet the citizens fared better than the losing side of most sieges. Mehmed was anxious to convert the city into the prosperous showpiece of his kingdom. He quickly reappointed the civil administrators. By noon the sultan had made his triumphant entry. He said his afternoon prayers at the majestic cathedral, now hastily transformed into a mosque.

The next day an eerie silence fell on the city. "Neither man nor beast nor fowl was heard to cry out or utter a sound," a chronicler noted. "The city was desolate, lying dead, naked, soundless, having neither form nor beauty."

Word of the fall of Constantinople reached Venice on June 29, Rome a week later. The Christian world was stunned. "The glory of the East," lamented Cardinal Bessarion, "the refuge of all good things, has been captured." It was one of the most ominous pieces of news ever to reach Europe. Gunpowder was beginning to change the world.

4

THE DEVILLS BIRDS

THE FRENCH KING Charles VIII was a fop. The grandson of the man whose guns had blasted the English off the Continent kept a royal perfumer to supply him with concoctions of orange blossom, civet, and oil of roses. His father, Louis XI, had died in 1483 when Charles was only 13. The young prince had left the administration of his realm in the hands of his older sister Anne, and had spent his time reading novels of chivalry. He developed a quixotic fascination with the glories of war. On grasping the reins of power in 1492, Charles realized that he had the means to turn his flamboyant dreams into reality.

During the previous half century, artillery had been undergoing a rapid transformation. Building on the foundation laid down by the brothers Bureau, French powdermakers and gun casters had reshaped the awkward bombard into an efficient prototype of the modern cannon.

Lighter, more maneuverable guns firing an energetic new form of powder created a terrifying destructive potential.

Charles, stupid and vainglorious, dazzled by the possibilities of the big guns, devised a plan that was as foolish in its conception as it was tragic in its repercussions. Claiming rights to the kingdom of Naples through a convoluted inheritance, he set off to conquer Italy. His ambition did not stop there. He planned to embark from Naples, retake Constantinople, turn back the Turks, and establish himself as emperor of the eastern world.

Though the Italy that Charles invaded in 1494 was that of Leonardo da Vinci and the Medici, it was the French dunce who held the trump card. His thirty-six heavy cannon, fashioned from gleaming bronze and firing compact iron balls, battered the fortifications of the Florentine stronghold Fivizzano with such violence that the citizens of many other Italian cities surrendered at the mere approach of the French batteries. The cannon were "planted against the walls with such speed," one observer reported, "the space between the shots was so brief, and the balls flew so speedily, and were driven with such force, that as much execution was inflicted in a few hours as used to be done in Italy over the same number of days." This was no exaggeration. In February of 1495 the French attacked the Neapolitan citadel of San Giovanni, a fort that had earlier withstood a siege of seven years. The cannon opened a breach in four hours.

The 24-year-old Frenchman won Naples, but with his lines of communication drawn thin and a league of Italian states and their allies forming against him, he headed back to France. Ironically, his heavy guns so slowed his retreat that he narrowly avoided being defeated by the Italians in the battle of Fornovo. Charles died three years later from an accident. He left a heinous legacy—Italy suffered more than sixty years of war as the French vied with the Hapsburg emperors and their Spanish allies for control of the peninsula. The Italian city-states, forced to make alliances of convenience with the invaders, saw their autonomy fade.

CHARLES VIII, for all his temerity, had taken advantage of a new potential of gunpowder. Throughout the previous 150 years, military strategists had been seeking a role for the dangerous black powder, and by the end of the fifteenth century they had achieved fundamental breakthroughs in powder formulation and gun design. Their achievement would constitute one of the key engineering feats of the Renaissance, bringing gunpowder technology to a level of development that, with minor changes, would remain the state of the art for a remarkable 300 years.

This feat was accomplished with scant contribution from science, which had yet to find a solid footing or rigorous method of experiment. Rather, the work was carried on by craftsmen who were described as masters of artillery, gun founders, powdermen, or gunners. They were in fact the earliest engineers. Engineering had originated as a military craft—engineers handled the "engines" of war such as catapults and trebuchets—and designed fortifications to defend against them. When cannon arrived, engineers became versed in the manufacture and use of gunpowder and in the forging of artillery.

Early cannon presented these engineers with both unprecedented opportunities and stubborn, complex problems. The most pressing difficulty was the inordinate danger that went hand-in-hand with gunpowder manufacture. Though gunpowder does not, like some modern explosives, detonate easily from impact, it is wildly sensitive to spark or flame. A teaspoon of gunpowder can dissipate its gases when it goes off, resulting in a harmless cloud of smoke. But a few pounds of loose powder generates so much hot gas so quickly that it can blow a building apart. Accidents were frequent.

In large part, the danger came from the demands of making effective powder. If gunners had loosely mixed the explosive's three ingredients they would have created a grayish powder that burned fitfully without exploding. The oxygen given off by heated saltpeter had to

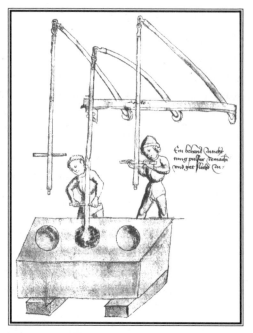

Grinding powder by hand in the fifteenth century using pestles attached to spring poles.

combine immediately with the sulfur–charcoal fuel in order for an explosive chain reaction to get going. For this to happen, the powdermaker had to force the ingredients into contact at the most minute, the most intimate level.

His means of doing this was to grind the ingredients in a mortar with a pestle, a laborious process known as "incorporation." Twenty-four hours of pounding was needed in order to bring out the explosive potential. The result was a material as fine as flour. Gunners called it "meal," or "serpentine" after a primitive artillery piece that used it.

During the incorporation, the powdermaker faced the greatest danger. Friction, a bit of iron that gave rise to a spark, or carelessness with a lamp meant instant cataclysm. When the fine powder was agitated it gave off a cloud of dust. This made the mill extraordinarily hazardous – the dust could drift to an open flame, ignite, and carry the fire back to the mass of powder. Dust could easily seep from cracks in barrels or waft from open containers, extending the danger to anyone who transported, stored or used the powder.

Another goblin that plagued early powdermakers was damp. Water molecules in the air cling to the surface of certain solids, build up, and gradually wet them. These substances are said to be hygroscopic. Ordinary table salt shows its hygroscopic nature when it clogs the shaker in humid weather. Charcoal is somewhat hygroscopic. It becomes more so when it's finely ground because of the greater surface exposed to moisture. When the moisture content of powder rises above one percent, the mixture begins to lose its explosive power. Stored in cellars, transported in rainstorms, shipped across the sea, gunpowder routinely deteriorated into damp, useless clumps. Saltpeter contaminated with calcium nitrate significantly exacerbated the problem. Gunners could never be sure whether their powder would provide a robust explosion or a disappointing fizzle. They spent a good part of their time drying and "repairing" powder.

Serpentine powder made the process of shooting early cannon slow and chancy. When gunpowder particles were packed tightly together, the mixture took fire only from the surface. The Chinese had made use of this quality when they developed their fire lances and rockets. But a gun required rapid deflagration of the whole mass of powder. Gunners could not completely fill the powder chambers of their bombards but needed to include carefully controlled empty space to assure that the powder was loosely packed and burned in an effective manner.

They loaded the powder chamber—the thick-walled container at the rear of the cannon—only about half full. They hammered in a wooden bung that sealed the chamber, then positioned the projectile just forward of it.

When fire was introduced to the chamber through the touchhole, the powder on the surface of the charge inside flared up. This combustion created turbulence that lofted additional powder, which instantly took fire as well. If the gunner packed the powder either too tightly or too loosely, or if he did not distribute it properly inside the chamber, the result was an erratic burn and a feeble shot.

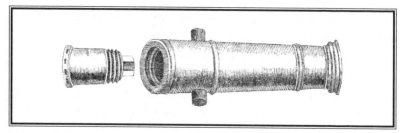

A bombard with its powder chamber unscrewed

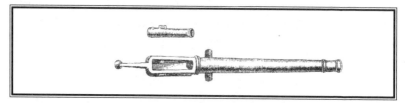

An early breech loading gun with separate powder chamber

Until late in the fifteenth century, loading a bombard took hours, and firing one could be an all-day event. At a siege of the Flemish city of Metz in 1437, a master gunner managed to fire three rounds from a large bombard in one day. The feat so astounded his employers that they made him undertake a penitential pilgrimage to Rome. Such rapid shooting, they assumed, could not have been achieved without help from Beelzebub.

One solution to the problems of damp and dustiness was to bring the gunpowder ingredients to the siege site separately and incorporate the powder on the spot. This eliminated the danger during transport, as the ingredients weren't explosive until combined. The gunners could also dry any damp saltpeter or charcoal before mixing. Such industrial operations on the plains of war were not unusual—blacksmiths and carpenters found similar employment.

At some point, probably during the late fourteenth century, powdermakers began to try a new approach to the incorporation process, one that would, over the century that followed, ameliorate both danger and spoilage and, as an unexpected bonus, enormously increase the explosive power and practicality of gunpowder. To reduce the danger during incorporation, engineers added a small quantity of liquid to the three ingredients before pounding them in the mortar. Sometimes they used distilled wine spirits, a substance thought to drive out impurities. Human urine was also favored, preferably of wine drinkers, ideally of imbibing bishops. The dampening reduced the dust and with it the risk of explosion. Instead of a loose powder, they ended up with a moist paste. Drying this paste was well worth the extra effort.

By adding liquid while incorporating, artisans also hit on a way to fend off the dampness the powder absorbed later. This counterintuitive discovery was related to the amount of surface area. A fine powder presents an enormous surface to the air, speeding the uptake of moisture. Instead of merely spreading the damp powder to dry, makers shaped the paste, known as mill cake, into balls. These might be of any size, from pebbles to chunks as big as loaves of bread. Once thoroughly dry, these hard masses minimized exposure to air because of their smaller surface. The result was a product that kept much better.

This single step, dampening the powder during milling, helped alleviate two key problems, safety and spoilage. But it raised another worrisome difficulty. Gunners carried to the war the gunpowder balls, known in German as *Knollen* or dumplings. They also brought along mortars and pestles, which they used, when the time came to fire the guns, to crush the balls to a powder. Inevitably, some of the explosive remained caked together as crumbs. This powder went off in the gun with much greater force. The increased brisance, as it was called, made firing a large gun even more of a death-defying act. Gunners tried to adjust by lowering the proportion of saltpeter in this powder, thereby diminishing its force. Still, the explosion could easily split wrought iron barrels, if not blow them to pieces. The powder chamber might disintegrate or fly off—this was this type of accident that killed the Scottish King James II.

Yet gunners were intrigued by the qualities of what they were calling "corned" powder, corn referring to a small grain. Some estimated that the new powder was 30 percent more powerful, others thought it two or three times as potent as serpentine. A cannon needed 34 pounds of serpentine powder to shoot a 47-pound ball, but only 18 pounds of the corned variety.

Today we have a pretty clear idea of why the crushed *Knollen* produced stronger powder. Fire spread through the mixture by means of a spray of hot, molten saltpeter and gas that leaped from a burning particle to its neighbors. This mechanism required space. To achieve a true chain reaction, particles of powder had to be surrounded by small gaps. Just as pieces of kindling are better than sawdust for starting a fire, gunpowder burned better in granular rather than floury form. The point was not to provide air—gunpowder supplied its own oxygen. But in the absence of gaps, the flame did not saturate the mixture. With corned powder, rapid combustion was assured.

The new powder also proved much easier to load into guns. Even when compressed tightly, the granules never fit together as closely as the tiny particles of serpentine powder. In effect, the new powder carried its own space, so gunners didn't have to leave part of the powder chamber empty. This simplified loading by letting the gunner pack the powder chamber, holding the charge in place with a wad of cloth rather than a wooden bung.

Gunpowder makers of the fifteenth century began to create these grains deliberately. They might have started by sifting the broken *Knollen* and selecting larger grains for their cannon. Eventually, they stopped forming the balls in the first place and forced moist powder through sieves, the size of the holes determining the dimensions of the grains. They varied the size depending on the type of gun in which they would use the powder. Large-grained powder seemed to work better in large cannon, finer grains in small arms.

Even as the corning process became standard, powder makers were learning to "purify" their saltpeter, a step that included converting calcium nitrate to potassium nitrate. The petermen took the liquor they leached from dung and mixed it with wood ashes, which contained abundant potassium carbonate. The calcium atoms joined the carbonate to form an insoluble material that dropped to the bottom; the nitrate radical was left to mate with the potassium. Artisans clarified this stew with the help of ox blood, alum, and slices of turnip. Relying on folk wisdom and crude experiment, they created the potassium nitrate needed for a durable and reliable powder.

All of these steps, discovered haphazardly over many decades, combined to yield a more powerful, more easily handled, and safer gunpowder. With only minor changes, corned powder would remain the standard ever after.

———

VANNOCCIO BIRINGUCCIO was 14 when Charles VIII's state-of-the-art guns brought war to the Italian peninsula in 1494. Born in Siena to a stone worker who served as superintendent of streets, the young man allied himself with the politically powerful Petrucci family and set out to become a contributor to the on-going improvement in gunpowder technology. The craft of fire was one of the most exciting professions open to a young man at the time, and one of the most modern. Encouraged by profligate military spending, such craftsmen had made enormous technical advances during the previous half century. Biringuccio would help push forward the allied fields of gunpowder, metallurgy, and artillery design.

Like the Renaissance artist who ground his own pigments and made his own brushes, Biringuccio was the embodiment of the integrated professional. He oversaw the entire process of gunpowder technology: assembled and refined the ingredients, ground the powder, supervised the mining and smelting of ores, designed and cast the guns. He arranged for the transportation of the bulky pieces to battle. During a siege he directed the positioning, loading, aiming, and firing of the big guns. If on the winning side in a conflict, he stage-managed the fireworks celebration that followed. Such a craftsman was in a perfect position to develop an intuition about gunpowder.

Yet engineers were operating largely in the dark when it came to theory. They had only a vague notion as to why the combination of ingredients in gunpowder produced an explosion. "Gunpowder is a corporeal and earthy thing composed of four elemental powers," Biringuccio speculated. "When the fire is introduced into the part of its greatest dryness by means of sulfur, it makes a great multiplication of air and fire."

Gunners had to be the masters of many trades. Mercklin Gast, a gunsmith from Frankfurt at the end of the fourteenth century, boasted that he was able to "restore spoilt gunpowder to its original state . . . separate and refine saltpeter . . . make powder that will last 60 years . . . shoot with large and small guns . . . cast from iron small-arms and other guns." Biringuccio developed similar talents including wire drawing, distillation, and the minting of coins. Their skills made them supremely practical men. Biringuccio had no patience with superstition. He scoffed at divining rods and the hocus-pocus of alchemists. "I have no knowledge," he wrote, "other than that gained through my own eyes." It was a startlingly modern perspective in the early years of the sixteenth century.

Like other craftsmen, most fire workers closely guarded the secrets of gunpowder formulation and cannon casting. They formed guilds, such as the *Fraternità di Santa Barbara*, which administered tests to hopeful apprentices, collected dues, allotted pensions. They passed knowledge of the trade only to trusted apprentices. In this, Biringuccio was an exception. He set down what he knew about gunpowder and metalwork in a book called *Pyrotechnia*. Written in the vernacular and published posthumously in 1540, less than a century after the invention of printing itself, the volume was the first printed book dealing with the gunner's various arts. It was a ground-breaking work, vastly expanding the practical knowledge available in this critical field. Nine editions were published over 138 years. By breaking the shroud of secrecy regarding the fire arts, it paved the way for treatises in many other practical fields. *Pyrotechnia* forms one of the deepest roots of the information age in which we now live.

———

ANOTHER ACHIEVEMENT of late fifteenth- and early sixteenth-century engineers that neatly complemented the new way of making powder was the improved design of guns. In all the millennia before the industrial revolution, metal presented artisans with staggering problems. Ores were elusive, mining techniques primitive. Extreme temperatures and complex chemical interactions were needed to obtain results. Something as subtle as the rate of cooling of a casting or the addition of a particular mineral salt to the smelting furnace could have significant impact on the final product.

Those who had grappled with the problem in the Middle Ages were the casters of the giant church bells that graced Gothic cathedrals. Bronze bells were the largest metal objects cast at the time—some weighed several tons. The resemblance between the gun and bell, both metal cylinders with hollow cores, did not escape the notice of gun makers. Forged wrought iron had offered a cheaper alternative to bronze casting. But the explosive force of corned powder called for a more durable container than the pieced-together bombard.

Since both contained large quantities of very valuable metal, bells and guns exchanged form with each other frequently over the centuries. During war, a city's conqueror invariably claimed its bells, melted them, and reshaped them into guns—a custom that continued into World War II when the Nazis looted thousands of bells from European churches. A peace treaty might see obsolete cannon re-formed into bells. In 1508 Michelangelo melted a great bell captured in Bologna to cast a statue of Pope Julius II. The Duke of Ferrara, known as *Il Bombardiere*, got hold of the sculpture three years later and destroyed it to cast a massive cannon that he named "Giulia."

Casting was always a demanding process. It required founders to achieve and maintain extremely high temperatures, to handle enormously heavy materials, and to bring about a precise mix of ingredients. Biringuccio, who discussed the casting of both bells and guns, commented on the difficulty of the art, which "appears to be more dependent on fortune than on ability." But as foundry expertise improved, and whenever cost allowed, rulers increasingly demanded stronger cast guns.

Gunners could not load these new guns from the breech. The powder chamber was formed as an integral part of the piece and had to be reached from the front. Corned power, though, made the process of muzzle loading straightforward. The gunner simply thrust a ladle attached to a pole down the barrel and upended it to insert a measured amount of explosive. He rammed a wad and ball in after it. Breech-loading guns faded and would not return until late in the nineteenth century.

In Burgundian-controlled Holland around 1450, gun makers began to cast lugs jutting from either side of the cannon barrel—together they formed a short axle. These "trunnions" significantly improved both mobility and aiming. Their secret was balance. Set just in front of the gun's center of gravity, the trunnions provided a pivot around which the gunner could raise or lower the barrel for aiming. They also added a support that allowed gunners to attach the barrel to a two-wheeled carriage—a modified cart of heavy timbers—without hindering vertical movement. The tail attached to the carriage allowed the gun to be swiveled easily from side to side. When the crew lifted the tail and attached it to a two-wheeled limber they created an articulated four-wheel wagon for transporting the piece. They no longer needed to dismount a cannon with hoists and set it on a specially prepared wooden shooting platform.

The successful manufacture of strong cast guns in turn led to denser ammunition. The carved stone balls that had been fired from bombards required the expensive labor of masons. When hurled against a wall, the ball shattered, dissipating some of its destructive energy. In searching for a better way to make use of stronger guns and powder, engineers found that cast iron shot, because it was three times as dense as stone, allowed a smaller gun to project as much force as a behemoth. Iron balls concentrated the force of corned gunpowder for greater effect. Because they could be produced using molds, iron balls obviated the masons' work. They could also be reused. And being more perfectly spherical, they fit into the cannon barrel with a smaller gap, making more efficient use of the force of the powder.

Renaissance soldier loading an arquebus

Engineers had assembled all the elements of a radical new technology: more explosive corned powder, smaller but stronger cast guns, dense iron ammunition. To make the guns safer while cutting down on their weight, they were cast with a thick breech end and a tapered barrel. Eight inches of solid metal at the breech would contain the sharp explosion of powder, while farther along the barrel, where the pressure was less, two or three inches might be enough. The barrels grew longer relative to the size of the bore in order to give the powder time to burn while the ball was still in the gun. The result was the classic tapered cannon that would set the standard for the rest of the gunpowder era.

The new guns, sleeker, longer, and lighter than the old bombards, threw their dense balls at a much higher speed. Gunners achieved velocity by gradually fueling their pieces with more powder. In early guns they limited the charge to 15 percent of the weight of the projectile. By the sixteenth century they were daring to load powder amounting to 50 percent, even 100 percent of projectile weight. The simple, easy-loading new cannon also made for a much more rapid rate of fire. At

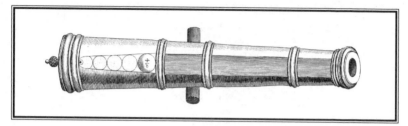

A classic cannon showing a powder chamber four times as long as the diameter of the bore

Brescia in 1564 an Italian gunner unloosed 108 balls from his heavy gun in five hours, a pace that would have astounded artillerymen of a generation earlier.

By the early 1500s engineers had devised a form of artillery that would be seen as the epitome of gunpowder weaponry for centuries to come. The cast smoothbore, muzzle-loading cannon was one of the most durable technologies ever invented. When armies fought each other in the American Civil War three hundred years later, they would field guns remarkably similar to those manufactured by Biringuccio and his fellows.

————

DURING THIS experimental period, the types of artillery pieces proliferated. Guns were named for birds of prey: Falcon and Saker and Sparrow-hawk.

The Basilisk took its name from the legendary serpent with lethal breath and gaze. Nightingale, one was called, Pelican.

Serpentine. Culverin, a snake, was applied to a whole class of guns.

Gog and Magog, they were called. The Doctor. The Pious One. Dragon, Lion, Fierce Buck, No-More-Words.

Brutal Butcher. "He will dance across moats, through ring walls, inner walls and bastions," its inscription said. "What he does not destroy directly will fall indirectly through his 'kiss.'"

In 1463 Louis XI christened two bombards Jason and Medea. Civic pride prompted names like Paris, La Dauphine, Londres, La

Plus du Monde. Strasbourg commissioned The Ostrich because its gunstones resembled enormous eggs. One big gun was called Lazy Servant Girl.

A cannon fashioned in 1404 for Sigismond of Austria warned: "I am named Katrin, beware of what I hold, I punish injustice."

Pope Pius II named two of his guns after himself, another for his mother. Emperor Charles V and the English king Henry VIII both ordered batteries to be known as The Twelve Apostles. In 1513 the cannon Saint John fell from grace when it became hopelessly stuck in the mud and was captured.

A poet complained:

> The devills birds I thinke were fitter names
>   To call them by, that spit such cruell flames.

————

WHILE CORNED powder was radically changing the nature of cannon, it was also elevating the handheld gun from a marginal to a central role as a way of throwing missiles. The earliest handgun had resembled a small cannon, with a stick inserted into a socket at the rear end to provide the shooter a better grip. This tiller also helped a soldier to aim the short tube—its appearance was not so different from that of a Chinese fire lance. The shooter ignited it the same way the cannoneer fired his piece, by applying a hot iron or slowly burning cord to the touchhole. An illustration from around 1400 shows a soldier with his hand cannon propped on a tripod.

Awkward to handle, these guns shared the problems of bombards when loaded with serpentine powder: If packed too tightly, the powder didn't have room to burn. The small chamber at the end of the tube made it difficult to insert just the right charge. Small arms also took time to reload, exposing the shooter to retaliation during a battle. As a result, though they appeared during the 1300s, they saw little development before corned gunpowder became common.

As the use of granulated powder spread, gun makers made the barrel longer and narrower. Like cannon designers, they used the new powder to fire a smaller projectile at a higher speed. But while the big guns stood on their own, handheld firearms had to be designed to fit a man's anatomy. Gunsmiths shaped the wooden stick so that the shooter could brace the piece against the shock of the recoil. At first the infantryman held the butt end against his chest. Later, gunmakers adapted the carved wood, known as a stock, to shoulder fire.

Applying fire to the touchhole while holding the gun steady was a challenge. The most dramatic improvement to the handgun came in the form of a lever attached to a metal hook. The hook held a smoldering piece of cord known as a match. When the shooter pressed the lever with his fingers, he lowered the match into a pan of priming powder beside the touchhole at the side of the gun, setting off the charge inside.

In the 1440s, a spring and sear (a catch for the lever) were invented. When the shooter pushed a button, the spring snapped the match into the pan. Soon after, a trigger replaced the button and has remained standard ever since. Because this mechanism resembled a household lock, it borrowed the name. The gun known as the matchlock took its place as the premier military firearm, beginning a reign that would not end until the flintlock arrived two centuries later. The handheld gun was now complete: lock, stock, and barrel.

This early matchlock, which became increasingly common during the late 1400s, went by the name "arquebus," a word that derived from the German *Hackenbusche* or hook gun. That name had been used for early defensive guns that included a hook to fit over the edge of a wall for steadying the gun and dampening the recoil. Powdermen made a special form of gunpowder for the weapon—finely grained, quick burning, fashioned to more exacting specifications, and more expensive than common cannon powder.

During the 1530s the arquebus was joined by a larger weapon, the musket. "Musket" was the term that later described all military in-

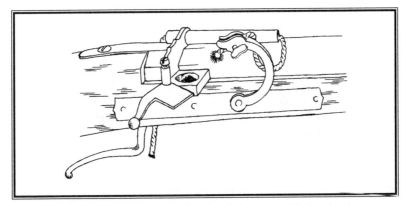

A sixteenth-century matchlock firearm. Pressing the lever upward brings the glowing match down onto the pan of priming powder.

fantry firearms, but this early version was a larger variation on the arquebus. It was intended to fire a ball that could penetrate the heaviest armor. Musketeers had to be particularly muscular to handle the powerful 25-pound gun, which they supported on a forked rest when firing. The heavy bullet, almost an inch across, blasted from the 54-inch barrel, could bring down a horse at 200 paces.

In spite of their simplicity, handguns required an elaborate loading routine. The shooter measured powder from a flask, poured it down the barrel, rammed wad and ball after it. He had to fill the primer pan with a finer powder from another flask. Then he blew on his length of slow match to make it glow and inserted it into the cocked lever. According to military historian Sir Charles Oman, "It was said that muskets would be more practical if Nature had endowed mankind with three hands instead of two." The chance of the gun going off when finally loaded was about fifty-fifty.

Matchlock weapons were far from perfect. Handling powder in close proximity to a burning match was a hazardous undertaking—a spark could spell disaster. Rain was another concern. A shower made firing impossible—the shooter could then only use the gun as a crude club. Yet after a century of experimentation, the handheld gun was evolving into a reliable means of killing from a distance. Like the

A musketeer firing at a knight

cannon, it arrived at a form that would not alter in any basic sense until the middle of the nineteenth century.

Through the work of the military engineers of the early modern period, gunpowder had finally come of age. These inveterate artisans had devised all the tools needed to turn the explosive powder into a truly lethal means of inflicting violence. Charles VIII took advantage of their labor to touch off a period of gunpowder violence that would disturb the peace of Europe for the next century and a half. During that time, gunpowder would increasingly dominate the plans of military strategists even as it assumed a troubled role in society as a whole.

5

VILLAINOUS SALTPETRE

THE FEUDAL SYSTEM had at its core the castle, a sanctuary that gave a local warlord his independence. The newly powerful guns solved a problem that had withstood many centuries of efforts by military engineers—how to breach a stone wall. Hammered with iron balls from the new cannon, the strongest walls could be reduced to rubble.

In response, defenders began to construct ramparts that were soft and low rather than high and hard. They made them from earth faced with brick, not stone. These new qualities allowed walls to absorb the energy of the cannonballs. To answer gun with gun, engineers designed walls as platforms for artillery. The cannon they mounted on them could destroy attackers' siege guns before they could be brought into position to damage walls. Drawing on mathematics and geometry, ancient sciences once again in vogue, Renaissance thinkers laid out "scientific" forts that enhanced the effectiveness of their defensive guns with carefully plotted angles of fire.

Some of the greatest minds of the Renaissance contributed to the frantic effort to counter the newly effective gunpowder weapons. Leonardo da Vinci, in spite of his conviction that war was *bestialissima pazzia,* worked as inspector of fortifications for Cesare Borgia. Michelangelo, who contributed to the design of the ramparts of Florence, wrote: "I don't know much about painting and sculpture, but I have gained great experience of fortifications." The German artist Albrecht Dürer, having studied in Italy, took the plans for this type of fort north. He published the first book on the new system, which spread across Europe under the name *trace italienne.* Over a period of just fifty years the forts neutralized much of the advantage of improved cannon and returned siege warfare to a new equilibrium.

Confronted with these fortifications, commanders looked for new ways to assault forts. They developed explosive mines, tunnels into which were packed quantities of gunpowder. Attackers in 1592 burrowed under the walls of a fortified bastion during a siege of the Flemish town Steenwijk. They exploded a large quantity of gunpowder in the mine, resulting in "the bodies of men . . . hovering piecemeal in the air, the torn and divided limbs yet retaining their decaying vigor and motion."

The petard was a squat iron pot, a cross between a gun and a bomb that held several pounds of gunpowder. A team of courageous and enterprising engineers would fasten it against a castle door, light a fuse, and run, hoping the blast would blow an opening. Shakespeare mentioned the device in *Hamlet,* relishing the irony of an engineer "hoist with his own petar." The derivation of "petard" from a common French word for fart would have elicited guffaws from the cheap seats.

Another weapon of the sieges was the mortar. Shaped like the apothecary vessel for which it was named, the short metal tube had a reduced-diameter powder chamber molded into the bottom. It used a small charge of gunpowder to heave a projectile upward, usually at a

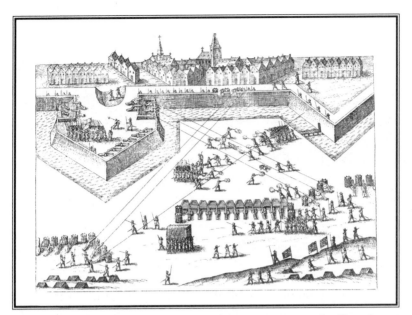

A fortification like those introduced in the sixteenth century to counter the effects of cannon

45° angle. Most often it threw exploding bombs—iron shells filled with gunpowder. Gunners did not dare to shoot these shells from longer guns for fear they would explode in the barrel. The stubby and more portable mortar, whose projectiles plunged onto the roofs of enemy fortifications from above, thus provided another dimension to gunpowder warfare.

The new forts were painfully expensive, the labor required to build them prodigious. Thirty million bricks went into the citadel of Antwerp, finished in 1571. The cost of the guns and powder to defend the walls was equally daunting. Unlike medieval castles, these fortifications were beyond the reach of petty lords. They were strategic, not tactical strongholds. They were the implements of kings and emperors, of centralized states with ample resources. Their appearance began to solidify the previously fluid borders of nations. "The modern frontiers of Europe," notes military historian John Keegan, "are, indeed, largely the outcome of fortress building."

LEONARDO DA VINCI imagined a large array of military implements, ranging from submarines to helicopters, that far outpaced the technical possibilities of his day. Around 1500, though, he sketched a design for an improvement on the handgun that would become a milestone in the history of firearms. Gone was the length of saltpeter-impregnated cord, a major drawback of matchlock weapons. An unreliable system of ignition to begin with, it was highly vulnerable to weather, and at night it gave away the shooter's position. Quartermasters had to keep an army constantly supplied with match in the face of the enemy.

To replace the matchlock, Leonardo invented a device, known as a wheellock, that resembled the works of a modern cigarette lighter. A serrated wheel extended into the pan of priming powder from below. The shooter wound this wheel against a powerful spring. Instead of a match, the lever hovering above the pan held a fragment of iron pyrite. When the shooter pulled the trigger, the pyrite dropped onto the wheel, which spun rapidly, generating sparks. The sparks ignited the powder, which carried the fire through the touchhole to set off the charge inside.

German gun makers adopted wheellocks enthusiastically during the first years of the sixteenth century. The mechanisms were delicate and expensive. No kingdom could afford to equip ordinary soldiers with costly wheellock weapons, but gunsmiths included the mechanism in expensive hunting rifles. They also applied them to short-barreled pistols, which were favored by cavalrymen. These guns were significant because they were the first firearms to include their own source of ignition.

By making guns easily concealable, the wheellock sparked social concerns that continue today. The first recorded firearms accident took place in Germany in 1515 when a man shot a prostitute in the chin while playing with a wheellock pistol—he had to pay her a pension for life. The use of wheellocks by highwaymen disturbed civil authorities and led to many edicts banning manufacture or possession of the weapons. In 1523 an ordinance in Ferrara outlawed wheellock weapons, "an especially dangerous kind of firearms . . . with which a homicide can easily be committed." Societies were beginning to sense a danger from the wider availability of gunpowder weapons, especially ones that an assassin could hide under a cloak. English authorities imposed an embargo on selling, firing, or making a pistol within two miles of Queen Elizabeth I.

The wheellock was only one way gunpowder's impact was beginning to be felt on and off the battlefield. By putting a new form of lethal power into the hands of commoners, gunpowder was among the elements that fertilized the long slow growth of feelings of rights and entitlements that would blossom into democracy. The idea of the individual firearm as an "equalizer," was not entirely fanciful. Gunpowder, as Thomas Carlyle would write in a later century, makes "all men tall."

Gunpowder accelerated a trend, introduced by bow weapons, that allowed killing at a distance. Hand-to-hand fighting, once the essence of war, became a less important part of combat. Missiles from both cannon and handguns took on a larger role in inflicting damage, breaking up formations of men, killing and wounding combatants. "It is rarely seen in our days," an English observer noted in 1598, "that men come often to hand blows as in old times they did."

Violence at a distance sometimes gave warfare a surreal aspect. At the 1582 siege of Oudenaarde near Brussels, Alexander Farnese, the duke of Parma, set up a table near the trench works and invited guests for open-air dining. The nineteenth-century historian J. L. Motley relates: "Hardly had the repast commenced when a ball came flying over the table, taking off the head of a young Walloon officer who was sitting near Parma. . . . A portion of his skull struck out the eye of another gentleman present. A second ball . . . destroyed two more guests as they sat at the banquet . . . the others all started to their feet, having little appetite left for their dinner. Alexander alone remained in his seat. . . . Quietly ordering the attendants to remove the dead bodies, and to bring a clean tablecloth, he insisted that his guests should resume their places."

Gunpowder also introduced a new element of mechanization into the behavior of armies. Battle increasingly called for cold rather than hot blood. Ferocity was not the quality needed in a man performing the intricate tasks of loading and firing a gunpowder weapon. The crews servicing artillery were even further removed from the role of the warrior. It was the gun that did the fighting—the gunners were, in the end, machine tenders.

Gunpowder made armed conflict inordinately expensive. Powder was dear enough itself. According to one estimate, each shot from a sixteenth-century cannon cost five thalers for powder and ammunition, an amount equivalent to a month's wages for an infantryman. Even more ruinous for the public treasury were the guns. The fabricators were highly paid specialists, the metal costly. And moving the guns to war required draft animals in unprecedented numbers. A fifty-gun Spanish artillery train operating in the Netherlands in 1554 used almost 5,000 horses.

The cost of gunpowder weaponry and its accouterments favored strong centralized states. Taxes became permanent and onerous. Cannon were, as one historian noted, the "ultimate tax-collectors." Smaller entities, such as dukedoms and independent city-states, were unable to afford an artillery train like that of Charles VIII and lost power to the princes and kings who could muster the resources. The foundations of modern nations were being built.

———

IN 1512 A SINGLE cannonball knocked down thirty-three men-at-arms at the battle of Ravenna, where Spanish and Italian forces were trying to fight off yet another French army bent on control of Italy. Gunpowder had made the battlefield a much more dangerous place. An observer at the siege of Maastricht said of the wounded: "Some lacked a leg, others an arm. Here there was a soldier whose guts were pouring from his body, and over there lay a man who had half his face torn away."

Ambroise Paré, born in 1517, saw gunpowder's effects at first hand. Apprenticed to a country barber, he had become a military surgeon and accompanied the French army on more than forty campaigns. Surgeons of his day were still of the artisan class, subordinate to the physicians who directed their work. Faced with the effects of gunpowder weapons, medical theories and practices were hopelessly lacking.

While swords, arrows, and battle axes had injured men grievously, the trauma inflicted by gunpowder was of a new variety. The heavy bullets fired by arquebus and musket—"leaden messengers," Shakespeare called them, "that ride upon the violent speed of fire"—fractured bones and burst internal organs. They carried dirt and bits of cloth inside the body, almost guaranteeing infection. Hits on limbs usually required amputation. Without anesthesia, the surgeon's speed was the only remedy for pain.

Because lead balls often flattened on impact, and because their force sent a shock wave through flesh, gunpowder wounds mystified medical men. Surgeons explained the extensive damage by theorizing that gunpowder left a residue of poison on the projectile that made the flesh rapidly deteriorate. They felt compelled to remove the poison from the patient's body—the preferred method was to cauterize the wound with boiling oil.

Treating a deluge of casualties during a 1536 assault, the young Paré ran out of hot oil. He found, not surprisingly, that his patients experienced considerably less inflammation and pain than those who received the standard treatment. The poisoned gunpowder theory gradually faded.

As gunpowder use spread, burns became a common injury of war. Guns malfunctioned; powder took fire accidentally in the chaos of battle. Intense pain, infection, and mutilation were the dire consequences. Early in his career, Paré had encountered three men during a campaign near Turin, "their faces wholly disfigured . . . their clothes did yet flame with the gunpowder which had burnt them." An old sergeant asked Paré if the surgeon could cure the men. He was told no.

The soldier "gently cut their throats without choler," Paré reported, rather than let them languish in misery.

The French surgeon was one of the first to note another deleterious effect of gunpowder. "One often sees gunners losing their hearing whilst drawing the machinery," he wrote, "because of the great agitation of the air inside the ear which breaks the aforementioned membrane." The delicate mechanism of the inner ear could not withstand the repeated concussion of cannon fire. Deafness was an all too common consequence of the gunner's trade.

In 1545 Paré wrote a best-selling text, "Methods of Treating Gunshot Wounds." Through close contact with the effects of guns, he came to execrate them. "We all of us rightfully curse the author of so pernicious an engine," he wrote.

———

IN 1498, A FLORENTINE war lord, Paolo Vitelli, put out the eyes and cut off the hands of enemy *schioppettieri*, because they had dared shoot their arquebuses at knights. European elites frequently deprecated gunpowder. By taking the muscle out of killing, powder favored the weak over the strong. It was the weapon of cowards. It killed by chance—the inaccuracy of all early firearms made this quite true. This randomness drained much of the perceived valor and meaning from fighting, much of its human quality.

Gunpowder was increasingly blamed for the death of chivalry, whose customs had already taken on the pleasant glow of nostalgia. "Through thee is martial glory lost," was how the Renaissance poet Ludovico Ariosto addressed the gun in 1532, "through thee the trade of arms becomes a worthless art." The repeated effort by European elites to blame gunpowder for the decline of chivalry was in part a smoke screen to obscure more profound changes in society that had long been undermining the role of the knight. Their objections were rooted in an understandable desire to maintain their monopoly on the use of force—gunpowder threatened to make violence too freely available. By the time handheld guns came into wide use, the objections were already growing hackneyed.

Another anti-gunpowder school denounced the sulfurous powder as a tool of the devil. "Some say the Poudre is the Meale of hell," a poet noted. Its invention was attributed to necromancers, practitioners of the dark arts. The stench, the grime, the association with witchcraft, all of these attributes gave credence to powder's diabolical connection. Ben Jonson referred to a mythical inventor "who from the Divel's-Arse did Guns beget."

John Milton carried this reasoning to its pinnacle in *Paradise Lost* (1667), relating how gunpowder had been fashioned by Lucifer in his primal war with the heavenly hosts. The satanic forces line up their guns and "those deep-throated Engines belcht, whose roar Embowell'd with outragous noise the Air." The archangel Rafael in vain warns Adam that humans must never invent such destructive engines.

While some churchmen cited gunpowder's infernal origin as a substantive argument against its use, prohibitions deterred few. The Papal States acquired guns as eagerly as any secular authority. The Church assigned gunners a patron, Saint Barbara. The exigencies of international competition were too pressing to be sidetracked by medieval theology. The idea of devilish cannon became a mere figure of speech.

Gunpowder figured in the Renaissance debate about whether all knowledge was the "rebirth" of ancient wisdom or whether the current age could make novel discoveries. The energetic new material served as convincing evidence for the modernists. Some writers asserted that the ancients must have devised gunpowder and used it as early as the siege of Troy. Where were the gun ports in ancient fortifications? skeptics asked. The progressive thinkers, for whom gunpowder was a symbol of the new age, won out.

Guns and gunpowder took on symbolic roles. The sexual connotations were obvious. Shakespeare wove a bawdy suggestiveness around a character in *Henry V*: "Pistol's cock is up, And flashing fire will follow."

The cupids engraved on powder flasks gave up their bows for small arms. The gun soon rivaled the sword as the prime emblem of masculine vigor.

By the end of the 1500s, those who scorned gunpowder weapons were seen as old-fashioned or hypocritical, the butt of humor. In *Henry IV, Part One,* the Bard has Hotspur express his droll contempt for a gun-shy courtier, attributing to the man the sentiment:

> . . . it was great pity, so it was
> This villainous saltpetre should be digg'd
> Out of the bowels of the harmless earth,
> Which many a good tall fellow had destroy'd
> So cowardly; and but for these vile guns
> > He would himself have been a soldier.

———

EVEN AS IT was roiling attitudes and upsetting conventions, gunpowder was delighting crowds of eager Renaissance spectators. Europe's earliest fireworks were shot off as a codicil to war. Gunners fired their pieces to mark the end of a successful siege or battle. They repeated the performance to give those back home a taste of the action, a whiff of powder smoke. During an era when all the arts were flourishing, the fireworks extravaganza became enormously popular. Bewitching shows originated in Siena and other Italian cities during the fifteenth century. Fire workers packed gunpowder compositions inside wooden or plaster figures used in dramas, so that they appeared to emit fire from eyes and mouths.

As gunpowder became more plentiful in the sixteenth century, fireworks spread across the continent. Gunners mounted a display in June of 1533 when Anne Boleyn, having recently become the second wife of Henry VIII, was crowned queen. It included "a great red dragon that spouted out wild fyre and round about were terrible monstrous and wild men casting fire and making hideous noise." All

A rocket-driven, fire-spewing dragon from a Renaissance pyrotechnic show

floated along on a barge in the grand procession to honor the royal couple. Perhaps the memory of these devilish, club-swinging ogres flashed through Anne's brain when, three years later, she watched the executioner step forth to remove her head.

Fashion gradually altered the form in which the fires of joy were displayed. In Italy, gunpowder effects became a feature of an artistic façade, known as a "temple" or "machine." The firemaster would direct the building of the temple, which might reach as high as a seven-story building, out of wood, canvas, and plaster. Artisans inserted colorful transparencies illuminated from behind, as well as statues, gilding, flowers, and painted murals. At night, hidden fountains of fire, rockets, roman candles, and other effects would erupt to captivate the onlookers.

In the Protestant north pyrotechnicians developed a more austere school of display. They made the fireworks themselves the center of attention, putting the actual devices on public view before shooting them. Motion was an important element of all shows, with figures driven along lines by attached rockets. Wheels spun furiously, flinging off sparks and fire—they were known as Catherine Wheels after a third-century saint who had been tortured on a circular rack.

Firemasters created their effects by varying the basic gunpowder mixture. Adjusting the proportions of charcoal or sulfur and adding additional materials allowed them to obtain a range of effects. They mixed coarsely ground charcoal with the powder when they wanted sparks that would float in the air—the charcoal did not burn away instantaneously. They found that metals, when reduced to a powder, were highly combustible. They included fine filings of iron or copper in a mixture to generate brighter light and more sparks.

Construction of firework devices demanded meticulous craftsmanship. A roman candle, for example, required the pyrotechnician first to make "stars," lumps of a gunpowder composition that he dampened and formed into hard balls that would burn for a second or two. Then he mixed up a second powder, called candle composition, this one heavy on charcoal and containing a little sugar. He would pour some pure gunpowder into the mouth of the tube, place a star on top of it, add a layer of candle composition. Then another small charge of powder, another star, more candle composition, continuing in this way until he had packed the tube to the top. When lit, the uppermost layer of candle composition burned vigorously, spraying sparks. When the flame reached the top layer of gunpowder, the small explosion threw the already burning star into the air. Each layer performed in the same way, sending up flaming balls one after another. Roman candles are made in exactly the same way today.

Styles continued to evolve. Dragons filled with various fire-spitting devices fought regularly against effigies of St. George. Ships bombarded each other to the accompaniment of cannon fire "as if the God of Battle had been there present." Casimir Simienowicz, firemaster for the king of Poland, advised in 1650 that the pyrotechnician might remind his prince of the "Incertitude of Prosperity, by a sight of the *Wheel of Fortune*."

Over time official fireworks displays grew in extravagance. The era of excess culminated in the celebration of the Peace of Aix-la-Chapelle, a treaty that temporarily ended hostilities across Europe in 1748. Fireworks marked the event in many capitals. In England, preparation went on from November until the following spring under the direction of Italian firemaster Gaetano Ruggieri. Military carpenters built a temple as high as a ten-story building and as long as two city blocks from timber covered with whitewashed canvas. Ruggieri arranged 10,650 rockets, shells, and pinwheels on this structure and began lighting them at 6 P.M. An orchestra accompanied by 100 brass cannon added to the excitement with a rendition of "Musick for the Royal Fireworks," composed for the occasion by George Frederick Handel.

But the handling of gunpowder remained tricky even for experts. The left wing of the temple caught fire and burned, slowing the performance. Ruggieri was still shooting his material at midnight—a considerable amount of it went unfired. Public disapproval of the waste led to simpler displays in future years. In any case, the English fared better the citizens of Paris, where a dispute over who was to light the city's magnificent display led to a brawl between French and Italian pyrotechnicians and an explosion that killed forty spectators.

———

GUNPOWDER, WHICH had seen its first duty as a species of battlefield theatrics, became a standard element in the theater. During the production of the play *Mystery of St. Martin* in 1496, an unfortunate actor in the role of Satan was awaiting his cue under a trapdoor. He was dressed in devilish regalia that, for authenticity, included charges of gunpowder. Just before the critical moment, "his costume caught fire round his buttocks so that he was badly burned," an account of the performance noted. The trouper was "so swiftly succored, stripped, and reclothed" that he was able to perform his part in the play without any sign of pain before going home to recover.

A century later Shakespeare and his colleagues eagerly included cannon fire in their productions. When the chorus in *Henry V* intones, "With linstock now the devilish cannon touches, And down does all before them," the stage directions call for "Alarum, and chambers go off." These special effects brought the sound and smell of gunpowder to audiences.

"Green men" became a standard of theatrical performances, parades, and outdoor spectacles. These enthusiasts covered their clothes with green ivy, wore black beards, and "were very ugly to behold." Pyrotechnicians made the spark-spewing fire clubs that the green men swung to clear the way for processions, dancing wildly and tossing firecrackers into the excited crowd.

Sometimes the effects went too far. A 1574 edict admonished impresarios for "sundry slaughters and mayhemmings of the Quenes Subjectes ... by engynes, weapons and powders used in the plaies." The warning didn't take. In June of 1613, three years before Shakespeare's death, the King's Players, enacting *Henry VIII* at London's Globe Theater, added realism by firing some gunpowder. Sparks landed on the thatched roof—the theater burned to the ground. Fortunately, everyone exited without harm. "Only one man had his breeches set on fire," an account noted, "that would perhaps have broiled him, if he had not by the benefit of a provident wit put it out with a bottle of ale."

6

CONQUEST'S CRIMSON WING

WHEN PORTUGUESE explorer Fernao Peres sailed his small fleet into Canton Harbor in 1517, he saluted the onlookers with a blast from his shipboard cannon. The concussions of the guns "shook the earth," echoed over the city, alarmed the locals, and drew angry protests from officials. As a Chinese historian later noted, "It had never before occurred to the Chinese that in some part of the earth a demonstration of war implements could also be an expression of respect or courteous recognition."

The incident embodied both a crucial misunderstanding and a deep historical irony. Gunpowder, invented five centuries earlier by the ancestors of those who crowded the Canton docks, had returned in a new and suddenly menacing form. The blasts of the cannon were emblematic. While the new explosive had seriously disturbed the

peace of Europe, its effects around the world would be even more jar-
ring and consequential. In a relatively short time, gunpowder would
radically alter relationships between Europeans and peoples in diverse
areas of the globe.

Before gunpowder's invention, conquest required sending contin-
gents of soldiers to exercise power over distant peoples. The conqueror
had to transport his troops, sustain them, and inspire them. Disease,
injury, fatigue, hunger, disloyalty, and the temptation of booty all limited the effectiveness of men's muscles as a tool of dominion.

Gunpowder's destructive energy could be contained in a wooden
cask. It needed no sustenance, was immune from illness, never mu-
tinied. The ambitious ruler could transport it great distances at little
cost. With the development of corned gunpowder and the improvements to cannon in the fifteenth century, a potent new means of extending political authority lay in the hands of European monarchs.

They faced one serious obstacle: cannon remained unwieldy instruments. The difficulty of moving them blunted gunpowder's effectiveness. Big guns sank into the mud, broke bridges, taxed the strength of
draft animals, slowed armies to a crawl.

The answer lay in a maritime revolution that was taking place in
northern and western Europe at almost the same time that gunpowder technology was reaching a new level of effectiveness. The sailing
ship was developing into an ideal vessel for transporting and using
guns effectively. Just as the adoption of gunpowder had introduced a
dramatic shift in mankind's conception of energy, the new sailing
ships overturned ancient notions of sea warfare.

The great sea powers of medieval times—Venice, Genoa, the Ottoman Turks—all relied on the galley as their principal fighting ship.
Driven by one or more banks of oars per side, each oar pulled by as
many as five rowers, these sleek vessels were fast and maneuverable,
the epitome of efficient muscle power. A commander would try to
drive the ram that jutted from the prow into an opposing vessel. Or he
would close with the enemy, allowing a contingent of soldiers to

swarm over the sides and engage in an armed melee. The violence was sharply concentrated. "Battles at sea," wrote medieval chronicler Jean Froissart, "are more dangerous and fiercer than battles by land, for on the sea there is no recoiling or fleeing."

Galleys had rudimentary sails, but rowers were essential for acceleration, maneuverability, and propelling the ship in calm conditions. A man pulling an oar generates only one-quarter horsepower of energy. Thus two hundred or more rowers were needed. Captains had to provide food and especially water, but the light ships had little cargo capacity. They could not remain at sea for more than a few days. The vessels needed relatively low sides to accommodate the oars, limiting their seaworthiness in storms. In the Mediterranean, where weather was predictable, tides minimal, and ports common, galleys reigned. Long voyages, large swells, and strong currents all challenged their capabilities.

Sailing ships reversed the advantages and drawbacks of galleys. High-sided and seaworthy, they were relatively difficult to maneuver, unable to accelerate on command, helpless when becalmed. As the dark ages waned, mariners had taken increasing advantage of sailing ships as trading vessels. They improved handling by replacing the steering oar with the sternpost rudder, which dropped straight down at the rear of the ship, and by adding more masts and refining the rigging of sails.

The improved ship offered an ideal platform for moving cannon— the ships' buoyancy countered the guns' enormous weight. But strategists quickly realized that sailing vessels could do more than transport ordnance. Ships could maneuver their batteries adroitly, bringing them to bear on a vessel or coastal position. The guns could give ships a new and devastating means of inflicting violence on the enemy.

Wind power meant that sailing ships could be handled by a small crew that was easily sustained. Their range was virtually unlimited. On arriving at their destination, the crew possessed a powerful tool for gaining an advantage over an opponent—they only needed to unpack the potential energy hidden inside their barrels of gunpowder.

THE MAN WHO found himself at the center of this dual energy revolution of sails and guns was the Portuguese captain Vasco da Gama. An intrepid mariner and a tough veteran of wars between his country and the kingdom of Castile, Da Gama was chosen by King Manuel I to command an exploratory journey to the East. He left Portugal in July of 1497 with four small ships, 170 men, and 20 cannon. He sailed down the coast of Africa, rounded the Cape of Good Hope, and continued into the unknown. His epic voyage, far more daring than that of Columbus five years earlier, brought him to the western coast of India in 1498.

The 38-year-old explorer was outclassed in the sophisticated trading world of the Indian Ocean. The gifts he brought to the ruler of the city of Calicut—some hats, six wash basins, two casks of honey—were insulting trifles, "the poorest merchant from Mecca gave more." No matter. Da Gama had found the route to the East. When he returned to India four years later, this time with ten ships, the casks in his hold contained not honey but gunpowder.

Da Gama had two motives for unleashing violence on his return. First, he had few other assets. Europe was still a backwater compared to the advanced societies of the East. The Indians had little desire for European goods, whereas European demand for spices made pepper the "black gold" of the day. One of the few exports that the Portuguese could offer was violence itself. It proved a formidable commodity. Da Gama's powerful gunpowder, iron cannon, and solid ships gave him an advantage over the meager guns and light vessels of the Indian Ocean traders.

The second stimulus was an irrational antagonism toward non-Christians, Moslems in particular. In part, this was a holdover from medieval thinking—Da Gama was a member of the military Order of Santiago, a knightly fraternity with roots in the Crusades. In part, the hostility was related to the very real threat that the successors of Mehmed II were presenting to Europe as they encroached on the eastern Mediterranean and moved inexorably through the Balkans. Da Gama's commercial considerations were reinforced by the goal that one conquistador described as "quenching the fire of the sect of Mahomet." A sixteenth-century diplomat summed up the mind-set of the conquistador: "Religion supplies the pretext and gold the motive."

The Indians had long possessed gunpowder technology. As early as the 1300s they had hired Turkish and European technicians to teach the fine points of grinding powder and shooting. Powder production was no problem, as India possessed the most abundant sources of saltpeter in the world. Yet the gun did not attract their focused attention as it did that of Europeans. The Portuguese astounded the natives with their guns, which fired "with a noise like thunder and a ball from one of them, after traversing a league, will break a castle of marble."

Da Gama wasted no time in putting his gunpowder superiority to use. He blasted stones at the recalcitrant Hindus of Calicut and set up a blockade to enforce his demands. He engaged in savagery that included burning women and children alive and sending a boatload of hacked-off heads and limbs into Calicut as a warning against resistance. Local powers were outraged. They assembled a flotilla of hundreds of ships and sent them out to attack the Christians. The battle that Da Gama was about to fight would offer a graphic illustration of how gunpowder had tipped the scales of power at sea.

––––

THE MARRIAGE OF gunpowder and ships, like many marriages, was both an ideal match and a relationship fraught with problems. Ships were the perfect way to transport and maneuver guns. But gunpowder carried a severe danger of fire and explosion, terrifying prospects on a vessel constructed of wood, pitch, and canvas. Controlling the violent recoil of the guns when they fired presented a particular challenge in the cramped interior of a ship. Aiming cannon on a heaving vessel was a chancy enterprise even for a skillful gunner.

Pieter Brueghel's depiction of a sixteenth-century war ship

Warriors had loaded guns on ships as far back as 1337, when the English sailing ship *All Hallows Cog* carried "a certain iron instrument for firing quarrels and lead pellets with powder." The earliest guns were mostly small pieces of ordnance fired to repel boarders or to harass the enemy at close range. Many were mounted on "castles," towering structures added to the bow and stern of ships to replicate the defensive advantage of castles on land—a high, protected position from which to fight. With the development of iron bombards, naval guns grew larger and better able to direct potent fire against enemy vessels.

Da Gama's crew watched the approaching Moslem fleet from over the rails of thick-sided, high-castled, and heavily gunned ships. The sound of gongs and war drums mounted to a pulsing clamor. His forces outnumbered, Da Gama quickly improvised a new tactic: He gave orders not to come to close quarter and not to board enemy vessels. He would fight a standoff battle. His ad hoc fighting method would mark the beginning of a new era of naval warfare, one in which gunpowder played the principal role.

The Portuguese captain was playing to his strengths. His largest ships each carried thirty-two sizable cannon, guns far more lethal than those of his adversaries. His crews stood ready with bags of pre-measured, corned gunpowder to reload the pieces. The massed Arab boats were an easy target. He brought his ships around so that their sides faced the approaching enemy.

The cacophony of war drums suddenly shrank to insignificance as the great guns roared. Stones tore into the light Moslem ships, often sinking them with a single shot. "It was not possible to miss," a participant reported.

The Portuguese forces won a decisive victory. Da Gama demanded that the owners of local trading ships purchase licenses, *cartazes*, if they did not want the same treatment that the Calicut fleet had just received. Forced trade became a standard practice—monopoly rights were granted to charter companies, who used gunpowder to assure their profits. The Europeans benefited from the divisions within the highly competitive trading community of southern Asia. Local potentates, hostile to their neighbors, curried favor with the intruders to gain an advantage. The Portuguese were able to maintain their lucrative monopoly with a force of fewer than 10,000 men.

In 1509 the Sultan of Egypt organized a formidable fleet of galleys in an attempt to reestablish Arab trading rights in the Indian Ocean. In a battle off the Indian port of Diu, Portuguese weaponry again proved its superiority. It was to be the last serious defiance of European hegemony in the region for a century. The next challenge to the Portuguese would come not from an Eastern power but from the encroachment of the Dutch, who could match them gun for gun.

———

As with artillery on land, the development of guns at sea required enthusiasts who were willing to bear the expense and supplant traditional methods of fighting. England's Henry VIII loved guns. He hired continental gun casters to set up operations in England and foreign naval architects to improve his warships.

Coming to the throne in 1509, the 18-year-old Henry took advantage of a simple but crucial innovation: the gun port. Until his time, commanders had positioned their guns on the open top deck to fire over the rails or "gunwales." The gun port was a hinged door built into the side of the ship that allowed cannon to fire from the lower decks. Their use meant that ships could carry more and heavier guns on decks just above the waterline. When the ship was sailing, the ports, closed and caulked, kept out the sea. During a fight, gunners swung the small doors open and fired out the side of the hull.

Ships changed in other ways. As the tactic of boarding enemy vessels faded, the castles shrank and disappeared. Once gunfire began to dominate, wooden castles offered inviting targets and scant protection. The term "forecastle" for the front part of a ship is the fossil of an obsolete design. Naval architects flattened decks and cleared encumbrances to give room for firing large cannon. They built ships wide at the water line to provide additional stability.

Henry also profited from technical advances that made it feasible to cast iron on a large scale. Much cheaper and more readily available than copper, iron had to be worked at the higher temperatures made possible by the blast furnace, which was then spreading through Europe. Gun casters designed the new cannon with thicker walls than bronze pieces to make up for the greater brittleness of cast iron. When it came to shipboard or coastal defense guns, the extra weight mattered little. Founders working in iron eschewed the ornate decorations that had become a hallmark of bronze guns. The new guns, painted black, were strictly utilitarian. The English began to fit out their warships with iron guns in 1534. They would soon develop a vigorous export trade in the less expensive weapons.

The development of the British navy under Henry VIII threatened the continental powers. The Spanish king Philip II was keeping a desperate hold on rebellious territories in the Netherlands even as he attempted to ward off the English pirates who were harassing the flow of goods from his new American empire. Forty years after the death of Henry, Philip hit on an immoderate solution: to invade England, now ruled by Henry's daughter Elizabeth I. He appointed the Duke of Medina Sidonia, a circumspect nobleman with little naval experience and grave doubts about the enterprise, as commander of a large fleet of warships.

In 1588 Medina Sidonia brought his "Invincible Armada," the greatest war fleet ever to set sail, to the English Channel to support the invasion. The English countered with warships that were the product of almost a century of development. The Spanish captains, though their ships were equipped with guns, still clung to the outdated tactic of closing with the enemy in the medieval manner. English commanders, with better guns and more maneuverable ships, fought the type of stand-off battle that Vasco da Gama had originated at the beginning of the century. Gunpowder, not ramming and boarding, decided the issue in favor of the English.

"Experience teacheth how sea-fights in these days come seldome to boarding," an English commission on reform noted a few years later, ". . . but are chiefly performed by the great artillery breaking down masts, yards, tearing, raking, and bilging the ships."

———

THE RIPPLES SET off by gunpowder's inception in medieval China continued to wash around the globe, affecting diverse societies in diverse ways. Like the Europeans, the Ottoman Turks saw in the explosive a tool that could help them realize their imperial aspirations. Mehmed II, known after his victory at Constantinople as

"The Conqueror," had been inspired by the success of his monster guns. During the second half of the 1400s he marched through the Balkans, capturing Serbia, Bosnia, and Albania and sending dismay through Christian kingdoms. Europeans, divided and with limited means for confronting the Turks' deadly cavalry, feared the worst.

Mehmed approached the forbidding problem of transporting his super-cannon by casting them on the spot, as he had done at Constantinople. For the siege of Rhodes in 1480, he hired gunners to make sixteen pieces, each 18 feet long and more than 2 feet in diameter. The Ottomans became famous for their giant bombards; a seventeenth-century European commentator marveled, "the Turks have such huge guns that they can tear down battlements only by their noise."

The guns' success seduced the Turks down the path of a dead-end technology. The gargantuan artillery pieces, as European powers had learned, were too slow and too heavy. Western armies were moving toward smaller, lighter guns fueled by fast-burning corned powder. The Ottomans, failing to grasp the advantage of the new guns, slipped behind in the arms race. Their momentum carried them to the gates of Vienna in 1529, but there they stalled.

In Japan, Lord Tokitaka had seen a Portuguese visitor fire an arquebus to knock a duck from the sky during the 1540s. He was so impressed that he handed over a small fortune in gold to buy the firearm. He ordered his expert swordsmith to copy the weapon. Gunpowder found fertile soil in Japan. The ceaseless wars among feudal lords encouraged craftsmen to continually improve the handheld firearm. They added an adjustable trigger pull and a lacquered box that protected the match and priming powder from rain. By the 1570s, these weapons had become an important part of Japanese arsenals. Lord Oda employed 10,000 arquebusiers who laid down a disciplined volley fire on the forces of a rival lord.

Yet with the arrival of the seventeenth century, the Japanese began to distance themselves from gunpowder. The government ordered both powdermakers and gunsmiths to clear their activities through a national commissioner of guns. Rather than advance the weapons' effectiveness, the shoguns stifled it. Over the next two centuries the use of powder dwindled until it virtually disappeared.

This long period of retrogression remains a curious eddy in the flow of history, one that raises fascinating questions about the idea of progress itself. To the Western mind, technical advances moved in one direction. The discovery of gunpowder was a momentous and irreversible milestone on the path of history, its increasingly effective use a foregone conclusion. European historians offered gunpowder as proof that Europe was immune from ever again being overtaken by the barbarism that had defeated the powderless Romans and Greeks. Gunpowder *was* civilization. For the Japanese, aesthetics, tradition, and politics trumped the advantages of gunpowder for more than two centuries. The Japanese would not turn to gunpowder again until well into the 1800s.

When Europeans arrived in the Americas, they came to a land where gunpowder was utterly unknown. Hernán Cortés described himself as "the instrument selected by Providence to scatter terror among the barbarian monarchs of the Western World and lay their empires to dust." The band of about 650 mariners and soldiers that he led from Cuba to the coast of Mexico in 1519 included thirteen arquebusiers. He also brought along ten heavy cannon and a supply of gunpowder. He put in charge of his artillery a man named Mesa, who had served as an engineer in the Italian wars. Mesa made effective use of his cannon when the Spaniards landed at Veracruz and the Aztec ruler Montezuma sent five emissaries to size them up. According to the Aztecs, Cortés had them bound and fired the "great Lombard gun" as a display. The shock of the sound, the Indians said, caused them to faint dead away. The emissaries took the alarming news to the capital. The very noise of the weapon weakened one, they reported. Showers of sparks belched from its mouth, along with fetid smoke.

The flash and roar of the cannon astounded the Indians, just as they had shocked early European troops. Again the guns' theatrical role supplemented their destructive power. In his 1521 siege of the

Mexican island capital Tenochtitlán, Cortés fired bombards from brigantines, demolishing the city building by building over a period of three months. The guns contributed to the demise of the Aztecs and helped bring about European domination of the New World.

At the end of the seventeenth century, the people of West Africa were being ravaged by the terrors of the Atlantic slave trade. One of the questions that mystified them was how the European traders—the red-skinned followers of Mwene Puto, Lord of the Dead—went about transforming human bodies into the trade goods that they brought back on their ships. The natives imagined that the white men, taking advantage of the fires of Hell that flared in the Land of the Dead, burned their black captives and ground their scorched bones into a powder. Packed in iron tubes, the black dust transformed itself again into fire and spewed pain and death whenever the violent and unpredictable people desired.

After textiles and liquor, gunpowder was the commodity most frequently bartered for human flesh. The Portuguese, concerned about proliferation of a dangerous technology, banned the importation of powder and muskets into Africa, but traders knew how valuable the commodities could be and smuggling was rife. The English and Dutch slavers had no compunctions about shipping gunpowder to Africa by the ton.

On yet another continent, gunpowder's magical connotations increased its impact. Firearms extended and reinforced a native ruler's supernatural powers. The intimidation of noise and smoke were as important as accurate fire. The growing dependence of tribal leaders on imported gunpowder further accelerated the practice of searching out captives and trading them to the foreign purveyors of death. The traders kept herding victims onto ships—their bones kept returning as barrels of the coveted explosive.

When the Portuguese arrived in China by ship in the early 1500s, they found that the Chinese had "some small iron guns, but none of bronze." "Their powder is bad," an observer noted. If the guns fired by the rude Portuguese offended the Chinese, such firepower could not fail to attract them as well. In 1522, government officials recruited two Chinese who had worked on Portuguese ships to explain the mystery. Later they turned to the Jesuits who had ventured east in search of souls. In the 1640s a German cleric built and operated a cannon foundry near the Imperial Palace. A generation later, Chinese officials imposed on Father Ferdinand Verbiest, a native of the southern Low Countries, to take over the operation. Verbiest protested that he was "little instructed in those affairs," but the emperor insisted. Taking information from books and passing it on to the workmen, Verbiest restored 300 old bombards and produced 132 smaller pieces. He solemnly blessed each gun and inscribed the names of saints and Christian symbols on the bronze barrels.

By the middle of the seventeenth century, the Chinese knew all the "secrets" of effective artillery. Yet neither their long history with gunpowder and metallurgy nor their proven ingenuity allowed them to match Western firepower. The reason is as elusive as it is historically intriguing.

If at times history is ruled by the authoritative voice of utility, it is at other times nudged forward by the whisper of taste, of fashion, of irrational whim. Chinese officials lacked the enthusiasm for guns that marked the leaders of European countries, from Edward III down to Napoleon Bonaparte. The denizens of the Chinese court looked on gunpowder technology as a low, noisy, dirty business. The fact that guns were useful did not matter, usefulness lacked the overriding value that it held for occidentals. What was more, the new cannon were foreign. To accept the ways of barbarians as superior and emulate them were deeply distasteful notions to Chinese mandarins.

The real reasons for the gap in gunpowder technology between Europe and the rest of the world remain both complex and obscure. When they fought the Europeans in the Opium War of 1841, the Chinese were still using Portuguese-made guns dating from 1627. In the end, a Chinese scholar was left asking, "Why are they small and yet strong? Why are we large and yet weak?"

———

ONCE THE STAND-OFF gunpowder fight became the standard way of delivering violence at sea, maritime battles changed little from one generation to the next. During the two hundred and fifty years that followed the Armada campaign, naval commanders directed artillery duels fought on wooden ships armed with smoothbore muzzle-loading cannon. Fighting in the era of the Napoleonic Wars differed only in detail from that in the days of Armada.

The mastery of gunpowder at sea required a man of extraordinary temperament and diverse skills. "A Gunner ought to be sober, wakeful, lusty, hardy, patient, and a quick-spirited man," a sixteenth-century writer advised. Three centuries later, Herman Melville, serving on a man-of-war, described the gunner firing his enormous artillery and, "with that booming thunder in his ears, and the smell of gunpowder in his hair, he retired to his hammock for the night. What dreams he must have had!"

Some of the nightmares that troubled the gunner's sleep involved the control of the violent recoil of his pieces. Simple mechanics dictated that the force pushing the gun backward was equivalent to the impulse that drove the cannonball ahead. On land, artillery captains allowed the piece to dissipate its force by lurching rearward. A ship offered little room for such a system. Gunners first let the ship itself absorb the gun's recoil. With a fight impending, they situated the gun jutting from its port and lashed it firmly to the ship's side. A man had to squeeze through the gun port and sit astride the barrel to reload. Jon Olafsson, an Icelandic gunner with a Danish fleet, performed this duty near Gibraltar in 1622. "The ship rolled all the starboard guns under, and me on my gun with them," he reported. "I swallowed much water and was nearly carried away."

A better solution was to allow the gun some recoil, but to restrict its range. Gunners designed a heavy oak framework that held the gun barrel by its trunnions. The weight of the timbers added to the inertia of the piece, absorbing some of the kick. Small wooden wheels on this carriage made the gun movable. A heavy breeching rope attached to the ribs of the ship on both sides and looped through a ring at the rear of the cannon brought the recoiling gun up short. With the gun thrust inboard, the crew could reload more handily. Once the piece was again ready to fire, the sailors heaved on tackle running through two sets of blocks that linked the gun to the side of the ship, hauling the muzzle out of the gun port for firing.

In spite of all precautions, there was an ever-present danger that a gun might break away from its breeching. The term "loose cannon" has become a commonplace. The reality of a three-ton mass of iron wheeling up and down the deck of a ship in heavy seas was truly terrifying. A runaway gun, "is a machine transformed into a monster," Victor Hugo wrote. "That short mass on wheels moves like a billiard-ball,. . . shoots like an arrow from one end of the vessel to the other, whirls around, slips away, dodges, rears, bangs, crashes, kills, exterminates."

If runaway cannon troubled the gunner's sleep, the danger of the volatile explosive in his charge engendered even starker visions of catastrophe. Warships carried several tons of gunpowder in their holds. A spark from two bits of metal clicking together could obliterate the vessel in an instant. The gunner required all fire on board to be extinguished before he supervised the loading of this massive quantity of explosive. He stored it in a magazine in the lowest part of the ship, where it would be most secure from enemy fire. He regularly turned the barrels over to counteract the tendency of the ingredients to form clumps. Because damp was a serious problem at sea, the gunner had to air his magazine regularly. If the ship was to lay over in a warm climate, he might even remove the entire store of powder to let it dry on land.

The magazine was the gunner's sanctum sanctorum. He secured it with a massive padlock—no one was to enter without permission from the captain himself. A marine sentry with a loaded musket stood guard over the precious cargo. This was not only a safety precaution. Mutiny was a very real danger on a vessel far from land and manned by impressed seamen subject to severe living conditions. Gunpowder represented the root of all power on the ship—a group of rebels who took over the magazine effectively gained control of the vessel.

The gunner coordinated his activities with a master-at-arms, who was in charge of training the men in shooting muskets, blunderbusses, and pistols. These small arms were distinguished from the "great guns" or artillery pieces. Marines—onboard soldiers—used the former to fire on enemy troops from perches on the masts, for repelling boarders, and occasionally for boarding an enemy vessel.

Guns on both land and sea gradually gave up their evocative names and were thereafter classified by the weight of the shot they fired. A 32-pounder, a gun used as the main battery of a ship, threw a cannonball of that weight, an iron sphere just over 6 inches in diameter. The gun itself weighed close to three tons. This mass of metal was needed to contain the blast of 10 pounds of gunpowder. The ball from this gun became a truly lethal instrument when directed against a wooden ship.

———

THE APPALLING intensity of sea battles can perhaps best be envisioned through the freshest eyes on board: those of the ship's boys. A large warship carried forty or fifty boys, about ten percent of the crew. A few were children of the wellborn, midshipmen learning to be naval officers. Most were delinquents or charity cases, poor urchins put to work at a young age. They were supposed to be at least thirteen years old, but many were ten or eleven, some as young as six. They performed menial tasks like cleaning the ship's pig pen, playing drums and fifes, acting as servants to officers.

During a fight, though, the boys were given a crucial assignment. They served as "powder monkeys," their job to hurry up and down from the magazine carrying the packages of explosive to the guns, most of which were arrayed on one or two enclosed decks below the open main deck. The danger of stray sparks meant they kept the cartridges either under their coats or in wooden or leather containers.

In the magazine, the gunner and his mates worked by illumination that seeped in from an adjoining light room, a closet containing lanterns whose light shined through thick bull's-eyes of glass. The gunner loaded his gunpowder into cartridges, sacks made of paper, silk, or flannel. The question of how much powder to use was the subject of intricate calculation and endless dispute among gunners. Manuals proposed all kinds of formulas. "Multiply the Weight of Ball by the Number of Diameters of the Chase in the Circumference of the Breech," one advised. "The Product multiply by 6, the last product divide by 96, the Quotient gives the Pounds required to charge the piece in action."

When a lookout spotted an enemy ship, taut expectation replaced the ordinary tedium of sailing. The crew rushed to prepare for battle. In minutes they cleared the decks, dismantling the partitions that normally formed officers' cabins to make room for working the cannon. Gun crews opened their ports and ran out the big guns, which were kept loaded at all times. They positioned buckets of water nearby, one for drinking, one for swabbing the barrel. They wetted the decks and sprinkled them with sand for better footing. They lit long lengths of match, readied flasks of priming powder. Rushing down to the magazine, the boys brought up their first cartridges. They stood by their assigned guns, giddy with excitement.

On the gundecks, which had barely enough headroom for a man to stand upright, the crewmen stripped to the waist to prepare for the ordeal. They tied handkerchiefs around their heads, partly to keep the sweat from their eyes, partly to muffle the roar of the guns. They unloosed the breech ropes, checked the tackle and other rigging of the monstrous engines in their charge. Then they waited, "all grim in lip and glistening in eye."

From the dimly lit deck, the crew caught only glimpses through the ports—now green water, now a far horizon, now blue sky, as the ship rolled with the swells. It was unlikely that anyone below deck would sight the enemy until a moment before the battle began. As smoke curled gently from the match tub, men turned pale, their throats clamped, their stomachs churned, their thoughts darted in a thousand directions at once. They yearned for action; they dreaded action. Then the action came.

The ship hove around. Towering white sails appeared against the blue. Below them, the menacing gape of rows of artillery pieces. Not always far off. British captains in particular were anxious to fight what were known as "yardarm actions," exchanges of fire from such close range that the yards of the two ships almost knocked together. Gunners sometimes could thrust their rammers out the port and touch the muzzles of the guns facing them.

Taking his direction from the captain, the lieutenant gave the order: "Fire!" The whole ship recoiled. "Every mast, rib and beam in her quakes in the thundering weight of the blow she has given." Cannon on land were loud—a row of guns firing simultaneously in the confined space of a ship produced a truly astounding roar. "My ears hummed," Melville wrote, "and all my bones danced in me with the reverberating din."

The boys were stunned by the sound. Their excitement transformed into wild feelings beyond words, they stepped forward, handed over their packages of gunpowder, and sprinted down to the magazine for more. Enormous clouds of sulfurous smoke surged from the mouths of the cannon. A large ship might burn half a ton of gunpowder a minute during a hot fight, clogging the decks with smoke.

The gun crew had little time to ponder the effects of their work. Their lives depended on speed. First a crewmember removed smoldering powder or debris from the last shot using a spiral of steel on a pole. Another seaman slid a wet sponge down the bore to extinguish any remaining embers. A third rammed home a cartridge of powder, followed by a cannonball and a wad of shredded hemp. A wire thrust through the vent pierced the bag of powder. The gun captain poured in fine priming powder from a flask.

Now came the heavy work of running the cannon out the gun port. A crew of eight men served the largest cannon. A 42-pounder with its carriage weighed 7,500 pounds, meaning that each man laying on the tackle had to repeatedly haul out almost half a ton of metal, an arduous chore, especially if the ship was heeled so that the slant was uphill. The work taxed the most muscular of mariners. During the wars of the Napoleonic era, the British government issued 40,000 trusses to sailors laid low by hernias.

With the gun run out, some of the crew had to hold it in position while the others adjusted the elevation, levering the breech end to the gun captain's instructions and inserting wedges to hold it. The fact that the gun's bore was not parallel to its tapered outer profile, combined with the ship's rolling motion, made fine aiming difficult. English gunners usually tried to skim balls across the water into an enemy's hull. Frenchmen often shot for their opponents' rigging, using special projectiles, split iron balls joined by chain or bar that went spinning through the shrouds and sails.

All these strenuous and dangerous tasks had to be repeated as quickly as possible—a crack squad might get off a shot every two minutes, even one a minute. The rate of fire was important, and the speed of gun crews was honed through endless drills. During battle, screaming officers urged them on.

The boys were an essential element of this assembly line of violence. Their legs ached with the constant running up and down, their ears rang painfully, their eyes burned with the acrid smoke.

Greater horrors awaited them. Like sharp echoes, the roaring of their own guns was repeated from the opposing ship. Balls flew past with a sound like ripping canvas. Or they slammed into the hull, great sledgehammers breaking open a wall. Now the boys saw why the decks and scuppers had been painted red: to obscure the splatter and flow of men's blood.

Chaos. A 14-year-old remembering an action called it "indescribably confused and horrible." In one sailor's words, "The very heavens were obscured by smoke, the air rent with the thundering noise, the sea all in a breach with the shot that fell, the ship even trembling, and we hearing everywhere the messengers of death flying."

Fist-sized balls and lacerating splinters whizzed past at invisible speed. "I was busily supplying powder," one combatant recalled, "when I saw blood suddenly fly from the arm of a man stationed at our gun. I saw nothing strike him; the effect alone was visible."

One man repeated the Lord's prayer over and over. Another reeled around in a kind of ecstasy, drunk on the intensity of the moment. Action left no room for sentimentality. "A man named Aldrich had one of his hands cut off by a shot," a sailor observed. "And almost at the same moment he received another shot which tore open his bowels in a terrible manner. As he fell two or three men caught him in their arms, and as he could not live, threw him overboard."

The boys enjoyed a moment's respite from this carnage when they ducked below for more powder. The area of the ship below the waterline was the only place secure from sudden death. Commanders knew how attractive the refuge could be to fainthearted crewmen; they posted guards with orders to fire on any man who tried to descend. The boys had to show their powder containers to gain admittance. If they took advantage of the privilege and tried to hide in the hold, sentries stood by to shoot them dead.

Picking up fresh cartridges, which resembled 10-pound sacks of flour, the boys returned to the awful scene they had just left. Sometimes they were too zealous. In 1761, on board the *Thunderer*, the powder boys in their enthusiasm brought up powder too quickly during a night fight. The pile of explosive went unnoticed in the dark. A spark touched it off. Thirty men died in the blast.

The boys ran along the deck, dodging the guns heaving backward in their recoil, careful to avoid the jet of fire that shot from the touchhole of each gun and scorched the beams overhead. They knew that they were hugging death. One boy had a spark reach his inflammable load: "His powder caught fire and burnt the flesh almost off his face," an observer noted. "In this pitiable situation, the agonized boy lifted up both hands, as if imploring relief, when a passing shot instantly cut him in two."

Sea battles are almost invariably wrapped in a cloak of glory. Horatio Nelson, who helped hone fighting tactics to a peak of brutality, now stands in state on his oversized pillar in Trafalgar Square. Yet few events, even in war, match the naval fight of the gunpowder era for sheer madness. That two bands of poor, illiterate, scurvy-ridden men, kidnapped and driven by the whip, should be induced to fire at each other from point-blank range with massive guns—it was a ritual of almost incomprehensible savagery and barbarism. That it should have continued and reached its apogee in the Age of Enlightenment is a deep paradox that any theory of political conflict is feeble to explain.

7

NITRO-AERIAL SPIRIT

THE QUESTION OF what the world is made of has always had at its core the mystery of fire. The fifth century B.C. Greek philosopher Heraclitus declared that the world was made of a single substance, "an every-living fire, kindling in measures and going out in measures." A century later, Aristotle ranked fire among the basic constituents of matter along with water, air, and earth. Anyone who thought about it took fire to be a thing, an element or entity, one of the building blocks of the world. Up until the seventeenth century, it had not occurred to philosophers that fire could be something quite different, a reaction, a process, a dynamic exchange between tiny, fundamental particles of matter.

Gunpowder played a key role in changing the way theorists thought about fire and about the nature of reality itself. It became one of the catalysts for the exciting tumult of ideas that swept Europe in the seventeenth century and developed into the modern conception of science. As the apotheosis of fire, gunpowder pointed to clues about the nature of this radiant phenomenon. Thinkers identified sulfur with the "sulfureous principle," the embodiment of combustibility. They knew that charcoal burned leaving barely any ash, suggesting that it was fire's ideal food. And saltpeter, or niter, which gave gunpowder its life, remained a wild card that any comprehensive theory would have to explain.

Chemistry, the science that would eventually reveal how the ingredients of powder worked, had yet to find a rational footing. The body of knowledge that attempted to answer questions about the substance of reality had no classical antecedents the way mathematics and astronomy did. Those who were trying to understand the material world could not look back to a Euclid or Ptolemy. Their predecessors were alchemists, sorcerers, druggists. The nascent field had no structure, no organized method. As a result, gunpowder, which represented mankind's furthest advance in the manipulation of natural materials, remained a deep mystery.

In Europe as in China, alchemists had developed the laboratory methods, procedures for purifying chemicals, and basic understanding of substances like saltpeter that contributed to gunpowder's early development. Later, their fanciful notions stood in the way of progress. Alchemists tried to understand the world in terms of resonances, correspondences, invisible connections between planets and metals, between the heavens and human life. Matter, to them, was still infused with the divine. Stars were alive. They made no distinction between natural philosophy and mysticism, between detailed observation and wild speculation.

In contrast to this holistic view was the deterministic philosophy handed down from Aristotle. His remarkably durable ideas, including the notion of a world made of four elements, still held authority in the time of Shakespeare. In the universities, Aristotle was revered as the fountain of all knowledge, even if his pagan philosophy didn't quite fit with the Christian conception of the universe.

The restless minds of the later Renaissance began to question both Aristotle and alchemy. Modern invention, which had given mankind the compass, the printing press, and gunpowder, as Francis Bacon noted, might well uncover wonders of which the ancients knew nothing. A new approach to knowledge began to emerge that was rooted in organized experiment. The theorists of the universities began to look at the techniques of gunpowder makers, where they saw phenomena for which they had no explanation. Their efforts to come to grips with the dynamics of gunpowder formed a bridge that would join technology and science, pushing the light of knowledge into the unknown.

———

BORN IN 1635, Robert Hooke was such a sickly boy that his father, a clergyman on the Isle of Wight, chose not to send him away to school. The lad read books, made clocks, experimented with homemade guns and gunpowder. After his father's death, Robert went to London at age 13, gave up on an apprenticeship as a portrait painter, and enrolled in the best school in England. Ten years later, at Oxford, he met Robert Boyle. Together, the two men would conduct experiments that would overthrow a view of reality that had stood for 2,000 years.

Boyle came from an entirely different social stratum than the man who went to work as his assistant. The son of an Irish nobleman, Boyle was enormously wealthy, well able to devote time to natural philosophy, which was for him an erudite hobby. Hooke, on the other hand, had to work for a living. His patron arranged for him to take a position as chief experimenter for the group of forward-looking London thinkers who would form the Royal Society, making Hooke the first person in history to earn his living from science.

Hooke and Boyle managed to devise a pump that could create a near vacuum in a bell jar. Inside this empty space a candle would not burn.

They found that when they used a magnifying glass to focus the sun's rays on sulfur in a vacuum, it fumed but would not catch fire. Clearly air played some role in combustion. Yet when the men dropped gunpowder onto a red-hot iron plate in a vacuum, it flared as usual. What explanation could they give? What did these experiments say about the nature of fire? What light did they shed on the mystery of gunpowder?

Boyle could formulate no satisfactory explanation. He concluded that niter gave off "agitated vapours which emulate air," but the observation was not supported by any known theory of matter. His book *The Sceptical Chymist* served as one of the founding documents in the new science of chemistry. But while he rejected the Aristotelian concept of the four elements, Boyle could not come up with a serviceable replacement. He left it to his one-time assistant to take the matter further.

Samuel Pepys noted in his diary that Robert Hooke "is the most, and promises the least, of any man in the world that I ever saw." As brilliant as he was emaciated, Hooke connected the experiments he had conducted with Boyle to two other phenomena. The first was an observation that went back at least to the Italian pyrotechnician Biringuccio: A metal when heated gained weight to form what was known as a calx. Lead, for example, gained almost ten percent of its original weight. The second was the fact that when Hooke pumped the air out of a container holding a mouse, the mouse died. The respiration of animals and the calcination of metals, he felt, had some link to combustion.

At the time no one knew what air was. Boyle had suggested that air contained "exhalations" that it picked up from the earth and from sunlight. Could these exhalations play some role in all three processes? To Hooke, gunpowder held the key. What the air contained, he decided, was a kind of niter that was essential for combustion, respiration, and calcination. Just as gunpowder contained sulfur, other bodies contained an essence of that mineral. "The dissolution of sulphureous bodies," he stated, "is made by a substance inherent, and mixt with the Air, that is like if not the very same, with that which is fixt in Salt-peter."

Hooke thus put forth the first coherent theory of combustion. Fire was caused by a substance in the air, a substance similar to the niter in gunpowder. Combustible materials contained a sulfur principle and air acted as a solvent. During burning, a portion of the flammable material was "dissolved and turned into the air, and made to fly up and down with it." The process created heat and smoke. "Fire is not an element," Hooke asserted in 1665. This was a revolutionary step forward.

John Mayow, a few years younger than Hooke, took a keen interest in his findings. After receiving an Oxford law degree, Mayow spent his life practicing medicine at Bath. He borrowed from Hooke in proposing that only a portion of the air was involved in combustion. Reasoning from experiments with gunpowder, he posited a "nitro-aerial spirit" that was contained both in the air and in saltpeter. When combustible or sulfurous particles collide with nitro-aerial particles, heat and light resulted, a flame. "The nitro-aerial spirit and sulphur are engaged in perpetual hostilities with each other," he wrote, "and indeed from their mutual struggle . . . all the changes of things seem to arise."

From the earliest musings on gunpowder in the 1200s, men had made the association between the boom of explosions and the crash of thunder. Gunpowder was thunder brought to earth. Now the idea had come full circle. Mayow's view solidified a folklore notion: Gunpowder didn't just mimic thunder; thunder was itself caused by a reaction between the essence of gunpowder's ingredients, niter and sulfur. Indeed, according to Mayow, all chemical reactions were a struggle between these two primal elements. As a physician, Mayow concluded that respiration was an absorption of the nitro-aerial spirit. He found that in a vacuum fresh blood effervesced but stale blood did not. He proposed the idea that muscular contraction resulted from a tiny "explosion of sulphur and nitre." The entire world, he felt, operated on a single dynamic—that of gunpowder.

Mayow's theory, particularly about meteorological phenomena, served as a magnet for commonplace observations. Didn't a hint of the sulfurous gunpowder smell linger in the air after a thunderstorm? Didn't niter, added to ice water, make it colder? Wasn't niter seen to act as a preservative for meat? Speculation took off. Surely the aerial niter in the clouds, with its "frigorific" quality, caused snow and hail. Niter's role as a fertilizer had long been known—you couldn't convince a farmer that a spring snow didn't increase the yield of his fields. And it made perfect sense that spirits of sulfur and niter, meeting under the earth, produced the violent explosions that surfaced as earthquakes and volcanoes. The sulfur associated with Vesuvius was ample proof of that.

Historians of science have pointed out the tantalizing fact that if Mayow had substituted "oxygen" for his "nitro-aerial spirit" he might have arrived at insights that would have leapfrogged chemistry a hundred years into the future. But Mayow died in 1679 at the age of 38, with science unable to discard completely the notion that fire was something hiding within a combustible substance.

The story shifted to Germany, where theorists turned back to Aristotelian categories. They posited an element that they held was responsible for combustion and that was found abundantly in living things. The irascible professor Ernst Stahl called this essence "phlogiston," from a Greek word meaning flammable. He used it as the basis of an all-encompassing theory of chemical reaction. He explained that when something burned it lost its phlogiston—it was "dephlogisticated." When a flaming candle was placed under a bell jar, the air inside became saturated with phlogiston and the candle went out. Combustion in a vacuum was impossible, for there was no air to absorb the phlogiston. According to Stahl, phlogiston was not fire itself, but "the matter and principle of fire." His theory, widely accepted for a century, promoted and studied by ardent "phlogistonists," was to become the last hurrah of fire as an essential element of nature.

———

DETAILED KNOWLEDGE of gunpowder's dynamics would have to await the slow development of chemical theory. Only in the nineteenth century would researchers bring its complex, lightning-fast, high-temperature reaction into focus. Because gunpowder faded from general use before the sophisticated instruments and techniques of the twentieth century fully developed, many details of its chemistry and combustion remain obscure even today. There has been no incentive to conduct in-depth research about a technology that is largely obsolete.

The studies carried out in the late 1800s found that 500 years of experiment by thousands of craftsmen had hit on a mixture—75 percent saltpeter, 15 percent charcoal, and 10 percent sulfur—that was close to the scientific ideal for the most powerful explosive. This mixture provided the amount of each ingredient needed for the most complete combustion.

Once ignited, powder burned at 2,138° centigrade. The intensity of the heat added to the explosive effect by causing the resulting gases to expand rapidly. The heat also created problems for gunners. Because the fire was hotter than the melting point of both bronze and iron, each successive blast inevitably wore down both the inside of the barrel and the touchhole. A series of shots in quick succession heated the gun dangerously.

The chemical reactions that constituted gunpowder combustion were complicated–they varied according to the exact makeup of the powder and the conditions under which the deflagration took place. In the simplest terms, the potassium nitrate reacted with carbon and sulfur to produce potassium sulfide, carbon dioxide gas, and nitrogen gas:

$$2KNO_3 + S + 3C \rightarrow K_2S + 3CO_2 + N_2$$

In reality, the reaction also produced other potassium compounds, carbon monoxide and traces of additional chemicals. The solids made up 56 percent of the results of the combustion, showing up as smoke and deposits on the inside of the gun barrel. The carbon dioxide, nitrogen and other gases constituted 44 percent of the products. These gases, at normal pressure and temperature, occupied 280 times the volume of the original powder. At the heat of the reaction, they took up fully 3,600 times as much space, generating a pressure of more than 20 tons per square inch in a closed vessel. This enormous pressure accomplished the explosive work. To get a sense of the scale of expanding gas, imagine a yardstick representing the solid powder stretching almost instantly to a length of two miles, representing the amount of gas generated.

The key to the reaction was speed. A lump of charcoal, because it was all fuel, actually released more energy on burning than did an equal amount of gunpowder, only a quarter of which was combustible. The natural burning of charcoal, though, gave off its heat over a much longer period of time. Gunpowder converted all its potential energy to hot, expanding gases in a few thousandths of a second.

In a cannon, a good portion of the chemical reaction took place before the ball had a chance to move. The hot gases acted as a powerful spring coiled between the 12-pound projectile and the breech of the gun. Because the ball was far lighter than the massive gun, it sprang forth with the greatest velocity, though the cannon received a formidable recoil jolt as well. Once the ball started moving, it traveled the length of the barrel in barely ten thousandths of a second—the blink of an eye lasts nine times as long. It gained all of its velocity during this brief acceleration. Once it emerged, along with a blast of expanding gas, smoke, and flame, it continued on its way with enough momentum to carry it a mile or more.

———

"THE FORCE OF gunpowder has hitherto served only for violent action," the Dutch theorist Christiaan Huygens wrote in 1673. "And although people have long hoped that one could moderate this great speed and impetuosity to apply it to other uses, no one, so far as I know, has succeeded."

Huygens was a genius in an age of geniuses and one of the first of a new breed of scientists. Growing up in The Hague of the 1630s, he lived near and probably met Rembrandt. Instructed by tutors, he learned to play the viola and the lute, to speak Greek and Italian. After attending the University of Leiden, Huygens plunged into an intensive investigation of natural philosophy. His interests were wide-ranging: He invented the pendulum clock and studied the rings of Saturn.

In the 1670s, Huygens turned to a problem that had stumped the best minds of his era. Mankind had used water and wind power with increasing effectiveness since medieval times. These proved useful when the work could be brought to the mill. But many tasks, particularly those associated with mining, required portable energy that could be brought to a specific location. The power of draft animals, the only alternative, was awkward and inefficient. How to devise a power source that could be located anywhere? It was a pivotal question.

Huygens imagined using gunpowder to drive an engine. He had only to witness the violence with which the explosive blasted a ball from a cannon to know that while powder gave rise to great energy, putting this energy to work directly presented a daunting challenge. He was familiar, though, with recent experiments showing that the air itself exerted considerable force when pressing on an evacuated container. Huygens thought this principle might offer a way to use the power of gunpowder indirectly. To accomplish his purpose, he turned to the cylinder and piston, long familiar as components of pumps. Huygens' flash of insight was to use the two elements not as a tool to move water, but as a source of power.

His *moteur à explosion* used a small charge of gunpowder to drive the air from the cylinder through a one-way valve. When the hot gases inside cooled, the atmosphere pressed on the piston, creating a power stroke. Only a small amount of powder was needed, so the force was easily manageable. Huygens speculated that with controlled gunpowder energy he could raise great stones for building, pump water, or turn mills. Unlike draft animals, he pointed out, the engine "requires no expenditure on maintenance while it is not in use." He calculated that a single pound of gunpowder could raise 3,000 pounds a distance of thirty feet. In a 1673 experiment he managed to lift a weight using a small cylinder.

Insurmountable problems kept Huygens from building a practical engine. The gunpowder explosion always left some gas inside the cylinder, sapping the device's efficiency. Even more critical, Huygens never hit on a way of delivering a succession of gunpowder charges into the cylinder.

Denis Papin, a Huguenot exile who served as Huygens' assistant, turned the idea in a new and more fruitful direction. Steam, he reasoned, might prove a more tractable means of driving an engine. Papin's idea prevailed—external combustion drove the industrial revolution. A return to internal combustion would wait another two centuries and the arrival of distilled petroleum as a fuel. Yet it is possible to envision the modern automobile driven by a row of cannon barrels whose explosive force is captured by reciprocating pistons. Though gunpowder itself never proved viable as a fuel for a mechanical engine, it played a decisive role in the birth of our most ubiquitous power source.

One of the effects of Huygens' work was to dispel the idea of gunpowder's infernal nature. Calvinist preachers praised him for attempting to divert the energy of gunpowder to peaceable applications. "The skillful achievements of chemistry are hated neither by God nor by Nature," one clergyman asserted.

The Dutch scientist was prescient in his understanding that the internal combustion engine would possess a high power-to-weight ratio, a quality the steam engine lacked. "Lightness is combined with power," he wrote. "This last characteristic is very important and by this means permits the discovery of new kinds of vehicles on land and water. And although it may sound contradictory, it seems not impossible to devise some vehicle to move through the air."

———

IN THE LATE 1600S, such prophetic speculation about the possibilities of gunpowder was leading thinkers to take a new interest in the ancient substance. Amateur scientists, inventors, and the plain curious sometimes traveled to view the rudimentary industrial enterprises where the powder was made. The mills—dirty, noisy, smelly, and dangerous—were hardly tourist attractions. But in 1673 a man named John Aubrey visited a powder works in Surrey, England. It was an impressive operation, he reported. Sixteen water wheels provided power for eighteen mills. The operation included "a nursery of earth for the making of saltpeter," a refinery for purifying this ingredient, and a corning house. The enterprise was "very well worth the seeing."

Aubrey was viewing one of the "new industries" that were beginning to lay the foundation for the coming industrial revolution. Artisans had long produced gunpowder, as they did most manufactured goods, in simple workshops. They set up their dangerous operations on the fringes of towns or cities, joining noisome industries like tanning or slaughtering. Not a few carried on the work in their homes.

Only gradually did the craft move toward a more industrial footing. The first water-driven English mill had been established in Surrey in 1555. A group of entrepreneurs erected five mills in the 1560s at great expense to themselves. In 1589, the year after the Armada scare, Queen Elizabeth tried to increase the nation's gunpowder self-sufficiency by instituting a formal monopoly on the part of licensed English producers. A limited number of substantial mills would provide a more reliable supply, it was thought, than many small workshops.

The recipient of the royal gunpowder patent was George Evelyn. Patriarch of a large family—he fathered twenty-four children—Evelyn and his descendants would dominate the powder trade well into the next century. The family continued the trend toward large, capital-intensive mills for making powder.

Obtaining sufficient saltpeter was still the key to a steady supply of powder. The Crown parceled to private interests, including the Evelyns, the right to collect saltpeter soil in the realm. Never satisfied with the yield, Elizabeth in 1561 paid a German named Gerrard Honrick £300 for a "statement of the true and perfect art of making saltpeter grow." As they did in other kingdoms, the English saltpetermen evoked the curses of the citizenry. Peasants had to be admonished against paving barns, a modernization that interfered with the accumulation and ripening of niter. A nation of bird fanciers greatly resented the intrusion of niter diggers into their dovecotes—in 1604 the Evelyns had to promise to replace all eggs and pigeons lost. The metamorphosis of the droppings of doves into the stuff of war was one of the wry ironies of the gunpowder trade.

During the 1600s, the industry began to outgrow the monopoly. The government was having difficulty containing unauthorized manufacture and trade. The importation of high-quality, inexpensive saltpeter from India spurred production. In 1627 the East India Company was allowed to grind powder from the saltpeter it was bringing from the East as ballast in returning ships. More mills opened. Powdermakers in Western counties supplied ships setting out from Bristol and Liverpool to trade in African slaves. Shipments to America picked up.

In addition to making new powder, mills were continually kept busy repairing or remaking gunpowder. Gunpowder was a perishable commodity in the best of circumstances, spoiled by dampness and clumping, or by agitation that turned grains to dust. The wooden barrels in which it was stored and shipped contributed to the problem. The saltpeter in the powder dried out the staves, opening seams and allowing moisture to penetrate. It was not uncommon that more powder was spoiled than was used. Sometimes the powderman could simply regrind the bad product, fashioning it again into useful grains. Other times he had to remove the saltpeter by dissolving it in water and start from scratch.

Danger remained the one constant of the gunpowder trade. The product could suddenly and without warning erupt into an explosive

mass of fire. The annals of gunpowder accidents are long and sometimes macabre. In 1647 a ship chandler in the middle of London was
packing powder into kegs. The cargo exploded, demolishing his house
and fifty others, including the Rose Tavern, which was crowded with
customers. Bodies were so mangled and torn apart that the number of
dead was impossible to calculate. The mistress of the tavern and a
bartender were found still at the bar, unmarked but stone dead. According to a *Gentleman's Magazine* report, a child in a cradle was
blown onto the top of a church, where she was found unharmed.

Because gunpowder was a critical military commodity, almost all
European governments either regulated or participated in powder
manufacture. France, for example, continued to rely on the medieval
craft tradition rather than moving toward the capitalist approach that
was taking over in England. Powder making was carried on in hundreds of small mills scattered around the French countryside. Powdermen worked like weavers or blacksmiths in factories that often
consisted of a single room. They might turn out half a ton of powder
in a year—the Evelyns provided as much as 250 tons to the English
crown.

Such a dispersed system required a strong central authority. The
power amassed by the Bourbon kings was backed up by a penetrating
administration that regulated the business down to the village level. In
1601 the king declared that the right to produce saltpeter or gunpowder was as sacrosanct as the right to coin money. All powder had to be
delivered to royal storehouses. Individuals who might want some for
hunting had to buy it from the government at fixed prices.

———

IN 1627, A MAN named Kaspar Weindl, who had marched to Italy
with the Austrian army, imagined he could use his knowledge of gunpowder to make money in the civilian world. In the Hungarian mining
region of Schemnitz, Weindl announced a new method for extracting

minerals from under the earth. He offered a demonstration for authorities. While they looked on, he packed gunpowder into a fissure in the rock, sealed it by hammering in a triangle of wood, and set off an explosion that fractured the stone. The mining tribunal was skeptical, but offered to let Weindl work in mine passages that had been abandoned because the rock proved too hard to crack with pick and chisel. Weindl was the first workmen ever referred to in records as a "blaster." His technique was the beginning of a whole new role for gunpowder in human affairs.

Before gunpowder, miners hammered and chipped at rocks, used picks and wedges, chisels and crowbars to break loose the ores. Alternatively, they lit fires against the rock, heated it, and dashed it with water to encourage cracking. The laborious work required both physical strength and a canny understanding of how to split solid rock. The prominent role that gunpowder later took in mining and tunneling raises the question, Why was it not used earlier? Why did three centuries pass after gunpowder began to be used in warfare before miners thought to put the explosive to peaceful purposes?

Gunpowder was a scarce and costly material during much of its early history. Mine managers had to show a significant increase in productivity to justify diverting powder from military needs. Nor were the miners themselves enthusiastic about adopting such radical technology. They preferred the ancient, established methods. To many, setting off explosions underground must have seemed foolhardy in the extreme.

Getting gunpowder to do the desired work also presented problems. Gunpowder only becomes explosive when contained long enough for the expanding gases to build pressure. The first blasters packed powder into natural crevices, then hammered in wooden wedges to seal the opening. The need to find substantial cracks and properly seal them limited this approach. Miners began to make their own openings, drilling a cavity into the rock by striking it repeatedly with a long chisel and mallet. Such work was tedious, time-consuming, and expensive.

Drilling, though, eventually became a miner's most important skill. One man, or sometimes a pair of workers, hammered at the drill, turning it with each blow, chipping a narrow hole into the rock. When the bore reached a depth of some three to four feet the miner packed it one-third full of powder, using about two pounds for the job. To contain the blast he inserted a cone of wood into the hole, the narrow end pointed outward, and filled the remaining space with powdered rock and dirt, leaving a space for the fuse. When the powder exploded, the immense pressure of hot gases trapped inside strained the rock beyond the point of fracture, shattering it.

It wasn't only in the depths of mines that engineers were turning gunpowder to civil uses. Blasting proved a valuable aid to quarrying and the building of canals. Because these waterways demanded a level and relatively straight path, obstructions had to be removed. To forge an alternate water route from the Atlantic to the Mediterranean, French builders spent two years blasting through Malpas Hill, outside of Béziers, in the 1690s. They carved a tunnel 515 feet long, 22 feet wide, and 27 feet high for the Canal du Midi. This was the first important canal tunnel and one of the first public works accomplished with gunpowder.

As the eighteenth century dawned, gunpowder was positioned to play an important part in the industrial revolution that lay just over the horizon. Manufactured on a large scale, and employed for the first time to further production rather than destruction, the explosive was taking on a new role, one that would eventually consume far more gunpowder than all wars combined.

8

NO ONE REASONS

"IT MAY WELL be called a roaring, nay a thundering sin of fire and brimstone, from which God hath so miraculously delivered us all," King James I proclaimed to the English Parliament on November 9, 1605.

Four days earlier, officials had uncovered the most audacious political conspiracy ever hatched in the scepter'd isle. The king and his family had been scheduled to attend the opening of Parliament, along with the lords, high magistrates, and bishops of England. A small group of men had schemed to explode two and a half tons of gunpowder under the Lords' chamber and with one blow decapitate the entire English government.

The energy concentrated in a keg of gunpowder made it a unique tool for bringing about political change. Before it became available, revolt required numbers. Assassination might erase a ruler but leave the government intact. An explosion that eliminated the principals of the state in a stroke and shocked an entire nation offered an irresistible lure to the dedicated revolutionist. The Powder Treason, one of the biggest news stories of Shakespeare's lifetime, became an early prototype of modern terrorism.

Addressing Parliament, the king labeled gunpowder a "most raging and merciless" weapon, which made no distinctions, offered no appeals. In this, he hit on the key horrors that have always accompanied the political use of explosives. They allow the perpetrator of the violence to distance himself from the scene of destruction, turning killing into a mechanical, inhuman act. They destroy indiscriminately anyone within range. This callous and arbitrary quality contributed a unique moral stench to such acts.

In the wee hours of November 5th, government officials discovered a "tall and desperate fellow" lurking in the shadows of a large storeroom under Parliament. They also found 36 barrels containing 3,600 pounds of gunpowder. By modern estimates, this was about five times what would have been needed to utterly demolish the building.

The man's name was Guy Fawkes. A muscular soldier of 35 with a bushy beard and a pious disposition, Fawkes was neither the originator of the plot nor its leader, yet for his involvement he was destined to be burned in effigy more often than any man in history. Having served in the wars, Fawkes knew about gunpowder. He had accumulated the mass of explosive and concealed it under heaps of firewood. With the touch of a match to a fuse, he planned to alter the history of the world.

The scheme was the brainchild of Robert Catesby, a patrician dabbler in violent fantasy. Known as Robin, Catesby was a wealthy 32-year-old widower, well-educated, six feet tall, shockingly handsome, with a charisma that entranced all who knew him. As uncompromising as he was eloquent, Catesby believed in his mission with such fervor that not even the Pope and bishops, who were urging restraint, could deter him. Catesby pulled together the band of conspirators, mostly relatives of his, and convinced them that "the nature of the disease required so sharp a remedy." After they swore on the Holy Sacrament to blow up the King and Parliament, they never wavered.

The "disease" was the prolonged and unrelenting persecution of English Catholics, part of the religious and political struggle that had begun when Henry VIII broke with Rome in the 1530s. His daughter Elizabeth I had banned priests from her realm and fined practicing Catholics, known as recusants, for failing to attend Church of England services. Catholics were not allowed to take degrees at universities and faced sharply curtailed career options. All through the late 1500s, though, believers remained firm in their hope. They hoped for a return to a Catholic monarchy. They hoped that foreign intervention might set things right in their country. They hoped that the death of Elizabeth, unmarried and childless, would bring about a new order.

Hope and despair are the hidden engines of political events. When a new ruler assumed the English throne in 1603, Catholic hopes seemed justified. James I, a plump, 37-year-old with a ruddy complexion, was the son of a Catholic mother, Mary Queen of Scots, and was married to a Catholic convert. He had made vague promises of leniency toward papists.

All hopes were soon dashed as James proved perfidious. Fines levied against recusants, briefly relaxed, were reinstated. A peace treaty between England and Spain extinguished dreams of foreign intervention. James's six children meant that England would not be at a loss for Protestant heirs. Despair at last gripped the Catholic faithful.

The plotters settled on their preposterous scheme in May of 1604. They acquired their first gunpowder that autumn. They were able to rent a room under the halls of Parliament in the old Westminster castle. Though referred to as a cellar, the space was at ground level and had convenient access to the Thames, facilitating the movement of powder. During the summer of 1605, the plotters stockpiled their barrels of explosive.

This new form of revolutionary violence was made possible by the increasingly widespread availability of gunpowder. In theory, the English government was maintaining a monopoly on powder production. In fact, merchants were eager to unload supplies before they deteriorated. The recent peace treaty with Spain had created a glut on the powder market.

There had always been private customers for gunpowder. Traders needed it to defend their ships on the lawless high seas. Privateers used it to attack enemy cargo vessels. In spite of official concern, gunpowder merchants handled the dangerous commodity with almost reckless insouciance. They were continually reprimanded for leaving stocks of the explosive unlocked. The Lieutenant of the Tower complained of one nearby dealer who was keeping 40 tons of powder on his premises. Curiously, English soldiers were required to pay for their own gunpowder. To defray part of the cost, some sold a portion on the black market. Anyone in the know could purchase a substantial amount of gunpowder with no questions asked.

However, as Parliament's long-delayed opening finally loomed, the conspiracy began to unravel. On October 26, Lord Monteagle, a Catholic nobleman and brother-in-law of one of the conspirators, received a letter. The writer advised him "to shift your attendance at this Parliament; for God and man hath concurred to punish the wickedness of this time." The conclave, the letter warned, would "receive a terrible blow."

Monteagle took the cryptic letter to government officials. Lord Salisbury, the secretary of state, put it before the king. This was more a concession to royal vanity than a sign of Salisbury's inability to penetrate the conspirators' intentions. James puzzled over the wording and finally struck on the notion that "blow" somehow referred to "blowing us all up, by powder."

The night before Parliament was to convene, the building was searched, the powder discovered, Guy Fawkes taken. Word of the deliverance leaked out immediately. Londoners rejoiced, lighting the first bonfires of what is still an annual celebration.

Robin Catesby fled. Having escaped to a safe house during a drenching rain, he and a handful of conspirators spread a small stock of gunpowder to dry before the fire. A spark touched it off. The violent burst of flame scorched Catesby and blinded another of the plotters, an ironic fate for men who had planned to shake the dome of heaven with this same explosive. The conspirators were duly rounded up, all but four who chose to shoot it out with the sheriff's men and were killed. Catesby died clutching an image of the Virgin Mary.

Guy Fawkes, interrogated by James himself, had the courage to face down a king—he refused to talk. James ordered that "the gentler Tortures are to be first used unto him and so by degrees proceeding to the worst." Two days of suffering were enough to wring a confession from the stalwart Fawkes. All hope for the remaining conspirators vanished.

Like other such plots, the Powder Treason produced effects that were the opposite of those intended. The barrels that had held the gunpowder were retained as relics to be "shown to the King and his posterity that they might not entertain the least thought of clemency toward the Catholic religion." Many English Catholics denied their faith in the period of intense prejudice that followed. Catholics would not be allowed to vote in Parliamentary elections until 1829.

Show trials convicted the plotters, along with the Superior of the Jesuits in England, who admitted learning of the plot in advance by hearing the confessions of two of the conspirators. All were given the treatment reserved for traitors. Each was dragged to the gallows in a litter, hung briefly by the neck, and cut down while still conscious. The executioner sliced off the condemned man's private parts before decapitating him and splitting his torso. On a dreary day in January, 1606, Guy Fawkes met this fate. Broken by his tortures, his joints painfully swollen, he needed help to climb the scaffold. As the noose tightened around his neck, he stepped off into history.

———

THE RELIGIOUS discord that engendered the Powder Treason was not limited to England. The Catholic–Protestant schism let loose antagonisms that shook kingdoms on the Continent as well. For several decades, Italy had been the area of Europe in which the air was almost perpetually tinged with gunpowder smoke. In the second half of the sixteenth century the violence moved north, to France and the Dutch Republic. It would soon spread to Germany, the wars there growing increasingly fierce. The core issues—nuances about the best way to achieve everlasting life through Jesus Christ—hardly seemed to justify mass slaughter, yet they led to an unimagined brutality.

Gunpowder added a horrific dimension to the wars of this period. Both handheld firearms and artillery were achieving a new efficiency, inflicting battlefields with storms of deadly missiles. "Firepower" became the military watchword of the day. Fueled by the lethal capabilities of gunpowder, violence raged across northern and central Europe until the middle of the seventeenth century, killing both soldiers and inhabitants by the tens of thousands.

Maurice of Nassau was 17 years old when he took power in Holland in 1584, assuming the title *stadtholder*. His father, William the Silent, had just been assassinated by agents of Philip II, the Hapsburg monarch of Spain. With the Dutch Republics struggling to win independence from Philip's empire, Maurice took a keen interest in military affairs. Commanders across the continent were searching for ways to make gunpowder the centerpiece of battle, not just an adjunct to it. Maurice and his cousins set up arrays of toy soldiers to explore new ways of using gunpowder weaponry. With the same fresh insight that had touched Joan of Arc during the 1420s, Maurice started a process of military realignment that would bring powder-based warfare to a new level of sophistication. The toy soldiers were a telling emblem—the fighting man was on his way to becoming a cog in the terrible engine of firepower.

Individual firearms had been around for a century, but no one had found a way to make them truly effective on the battlefield. Maurice saw that musket-wielding soldiers, extended in a line, could unleash a wall of fire to fend off and disrupt enemy forces. What he needed was a system to coordinate that fire. A student of the classics, he found the answer in the military thought of the Romans, the last power to rule Europe with an infantry-based army.

His goal was concentration and coordination of firepower. The secret for achieving it was drill, a means of making every soldier move in lockstep with his fellows. The formation that Maurice devised was a line of infantrymen ten deep. Those in the front row fired, then turned and countermarched to the rear, where they could reload in relative safety. Those in the next row stepped forward and fired in turn. A murderous ballet ensued, groups of men moving in unison and in close coordination with other units. Maurice broke down the process of loading a firearm into forty-two small gestures, each with a name. His troops practiced the movements over and over until they could be performed without thinking under the stress of battle.

Combined with a system of rigid discipline, drill melded the mass of warriors into a single unit. "No one reasons, everyone executes," is how the Prussian monarch Frederick the Great would describe it in the eighteenth century. Drill not only turned individual soldiers into an effective means of delivering violence with gunpowder, it also trained them to stand up to the harrowing return fire of the enemy. It gave soldiers the ability to perform a complicated choreography in the mouth of hell. The goal was efficient repetition and massed fire, not heroics. Discipline replaced initiative in war, a transformation that would be paralleled by the coming encroachment of factory production on the methods of the traditional craftsman.

A year after Maurice rose to the head of Holland, gunpowder trumpeted its role in the coming cataclysm in yet another way. Spanish Hapsburg troops under the Duke of Parma were besieging the Dutch port of Antwerp. An itinerant Italian military engineer named Federigo Giambelli had offered his services to the Spanish and been rebuffed. Like the enterprising engineer Urban at Constantinople, Giambelli gained revenge by peddling his skills to the Dutch.

Giambelli turned a sailing vessel, ironically named the *Hope,* into a new weapon: the first floating time bomb. He packed its hull with almost four tons of gunpowder and surrounded the explosive with bricks, scraps of metal, even tombstones. This debris would turn into deadly missiles when the powder went off. A clockwork device controlled the fuse. The ship was known as an "infernal machine," a term that straddled two worldviews: the medieval, now fading, of demonic influences; and the modern, of a mechanical, clockwork universe.

The Dutch set the *Hope* adrift on the tide. The vessel approached a densely manned pontoon bridge by which the Spanish were blocking access to the city. The bomb exploded on schedule, blowing a gap in the bridge and strewing wreckage for a mile in all directions. It was, to that time, the most lethal blast of a single weapon in history. Hundreds of men were killed instantly. The "Hell-burner of Antwerp," as it was called, the first major bombing in Europe, was a chilling portent of gunpowder's increasingly destructive potential.

The wars over religion, property, and empire reached their culmination in the Thirty Years War, a conflict in which France, Sweden, and the Dutch Republics battled against Spain, Austria, and Bavaria over Hapsburg supremacy in the Germanic lands. Marked by convoluted alliances and motivations, it was a "war for all reasons."

During this period of perpetual ferment, the man with the clearest vision of gunpowder's deadly future was Gustavus Adolphus, the king of Sweden. Gustavus ruled a rustic kingdom barely touched by the sweeping changes of the Renaissance. Yet the young king had a vision and energy that would, for a time, make Sweden a power to be reckoned with. An affable man with golden hair and myopic blue eyes, Gustavus insisted on sharing the hardships of his troops on campaign, even helping dig earthworks when needed. He was a fighting king, a consummate man of action. Napoleon judged him among the half-dozen greatest commanders in history.

Gustavus took the system sketched out by Maurice and brought it to fruition. He drilled his men incessantly and enforced iron discipline. He was determined to maximize their firepower. To that end, he issued lighter muskets and introduced paper cartridges, containers of

premeasured powder that allowed soldiers to load faster. Faster loading meant more frequent volleys of fire.

Gustavus, an expert gunner himself, introduced his most far-
reaching changes in the realm of artillery. Until he took power in
1611, the big guns had been used primarily for siege work, for war at
sea, and for the static defense of forts. The battlefield role envisioned
by Edward III at Crécy had continued to elude commanders. The
great guns, concentrated under the supervision of contracted gunners
rather than soldiers, remained ponderous and largely immobile during
the fighting.

Gustavus integrated the guns more skillfully into his forces, creating the first effective field artillery. He assigned a handful of smaller,
much lighter pieces to accompany infantry and cavalry regiments. If
cartridges could ease loading for musketeers they could do the same
for artillerymen—Gustavus ordered gunners to use prefilled bags of
powder with balls already attached. Because of these changes artillery
fire could be employed in the middle of a battle. The great guns, designed to smash stone walls but directed now against human flesh,
added to warfare a new dimension of horror.

By 1632, the war had already been raging up and down Germany
for fourteen years. Gustavus found himself in command of a massive
anti-Hapsburg coalition. He maneuvered a force of some 20,000 men
into position to attack an army of similar size led by Albrecht Wallenstein, the mercenary commander of the imperialist force, outside the
village of Lützen, fifteen miles west of Leipzig. Gustavus planned to
attack at dawn, but the dank November morning brought a thick fog
that stalled operations and gave Wallenstein time to recall a large detachment of cavalry.

The battle that ensued in many ways typified the war as a whole. It
was a story of firepower gone mad. The Swedes pushed forward into
withering musket and artillery fire to capture imperial cannon, which
they turned against the enemy at close range. But by doing so they lost
touch with their own cavalry, allowing the imperialists to regain momentum. The thick mist descended again, mixing with the gunpowder

smoke to blind the fighters and turn the field into a snarl of confusion. The effects of drill, rapid musket fire, and effective field artillery combined to inflict enormous casualties on both sides.

Though he was the champion of a cool-headed, almost mechanical approach to military strategy, Gustavus could not keep himself from personally leading a detachment of cavalry to reinforce a weak spot in his line. He was hit by a bullet. His horse carried him away from his escorts into the wild melee. An imperialist cavalryman shot him in the back. He fell. Another enemy soldier fired a lead ball through his head. Looters stripped him to his shirt. His horse careered riderless through the chaos.

The Swedish forces "won" the battle of Lützen, driving Wallenstein's army from the field. Yet the awful carnage and the loss of their leader made the victory less than sweet. Wallenstein, broken by the battle, tried to sell out the Hapsburgs and was assassinated. Lacking a conclusion, the war continued for another sixteen years. Finally, Gustavus's daughter Queen Christina joined Louis XIV of France as co-guarantor of the 1648 Peace of Westphalia, which was to bring order, if not peace, to the European continent until the French Revolution.

Gunpowder had, for the moment, turned military conflict into savagery. The ethos of the hunt had invaded warfare. Equipped with firearms, soldiers had turned predatory, chasing down beaten foes, killing prisoners, ravaging the countryside. Modern estimates put Germany's losses during the Thirty Years War at almost 8 million persons, more than a third of its population. The conflict had imposed on an entire generation a level of atrocity and degradation that horrified thoughtful observers. Gunpowder, which philosophers had once imagined would protect Europe from a slide back to barbarism, threatened to plunge the continent into a new dark age.

9

WHAT VICTORY COSTS

EVEN AS GUNPOWDER was ratcheting the violence of organized warfare to unprecedented levels, it continued to offer thinkers perplexing and intriguing problems. "Ultimately it was the effects of gunpowder on science rather than on warfare that were to have the greatest influence in bringing about the Machine Age," wrote historian J. D. Bernal. "Gunpowder and cannon not only blew up the medieval world economically and politically; they were major forces in destroying its system of ideas."

Beginning in the sixteenth century, gunpowder began to focus the attention of Europe's natural philosophers not only on the mystery of fire and the composition of the material world, but also on questions of mechanics and the forces and laws of motion. For example, both gunners and natural philosophers wanted to know: What happens to the cannonball after it leaves the barrel of the gun? The search for a definitive answer took four hundred years and required the creation of entirely new fields of science.

Gunners did have some notion about what drove the ball out of the gun. Vannoccio Biringuccio summed it up in 1540: Fire, he thought, took up ten times as much room as air, air ten times as much as water, water ten times as much as earth. So when the earthy powder changed to fire, air, and moist smoke, these elements instantly expanded, pushing the projectile before them. While based on a fanciful theory, his idea roughly summed up what was happening. Burning powder changed to a hot gas of vastly larger volume. The gas exerted pressure on the ball.

But why did the ball keep moving after it emerged? What path did it follow? And what did that path reveal about the forces acting on the projectile and on all objects? The first man to address these questions was a contemporary of Biringuccio named Niccolò Tartaglia. Niccolò was born in the northern Italian town of Brescia in 1500, the son of a mail courier. His father died when the boy was six, leaving his family in poverty even as wars ripped across the country. At the age of twelve Niccolò was caught up amidst rampaging French troops. A soldier slashed his sword across the boy's face, lacerating his mouth and palate. Niccolò's mother nursed him back to health, but he carried through life a severe disfigurement and speech defect. He gave himself the name Tartaglia, from the Italian word for stutter. His original surname is lost to history.

Once healed, the teenager went to Master Francesco to begin learning the alphabet. He had only reached the letter K before his meager funds ran out. He achieved all the rest of his learning on his own, "accompanied," he later wrote, "by the daughter of poverty that is called industry." He found in himself an affinity for mathematics and was soon in Verona teaching students to use the abacus. Later he became a professor of math at Venice, still barely making enough to support his family.

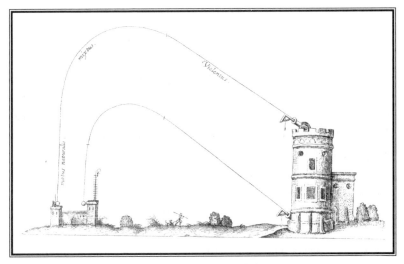

Seventeenth-century ideas about the motion of cannonballs

Until 1531, Tartaglia understandably took little interest in the implements of war. But that year, a gunner asked him at what angle one should aim a cannon to achieve the longest range. The question intrigued the young math teacher. He saw in it an opportunity to apply mathematical principles to a real-world phenomenon. He devoted a great deal of thought and research to calculating trajectories. He decided that an elevation of forty-five degrees gave the longest shot, which, in a vacuum, was true. During his research he invented the gunner's quadrant, a kind of carpenter's square equipped with a pendulum at the angle. When one leg was inserted into the barrel of the gun, the hanging weight indicated the degree of elevation. This device, along with the calipers, gauges, and levels used by gunners, helped introduced the notion of accurate instrumentation into science.

A ball fired from a cannon was invisible because of its speed. Until Tartaglia's investigations, gunners thought their projectile flew in a straight line until the end of its flight, at which point it simply dropped to the ground. They understood movement through the eyes of Aristotle, who had declared that there were two different types of motion. Natural motion—the fall of an apple, the rising of smoke—resulted from the tendency of all things to return to their proper places. Fire rose, earth sank. Violent motion, on the other hand, contradicted natural motion—an arrow shot into the air rose against its natural tendency to drop. This type of movement required a force acting on the object at all times—but what drove the arrow once it left the bowstring? Aristotle said the propelling force was air, which rushed around the projectile and pushed it from behind. The concept of inertia, like that of gravity, is familiar today, but in the 1500s the reason an object continued to move or fell to earth remained a profound mystery.

Tartaglia declared that "a piece of artillery cannot shoot one pace in a straight line." The greater the velocity of the ball, the flatter the trajectory, he admitted, but natural motion began to bend the ball's path as soon as it left the gun. Asserting that the projectile's course was at every point a curve, Tartaglia took an important step toward understanding the phenomenon of motion. He argued that the trajectory was determined by a struggle that pit the speed at which the ball was thrown forward against the force, whatever it might be, that drew it toward the ground.

On the basis of his theories, he drew up tables of distances that a gun would shoot at different angles and charges. While these charts were not entirely accurate, gunners continued to use them well into the seventeenth century. Tartaglia affirmed that mathematical science was needed to calculate "the strange varietie of the Circuites of all Bullets in the Ayre." He experimented with "gunpowder of divers sorts," investigated the weight and diameter of projectiles, and tried with little success to calculate the ideal length of a gun barrel.

Tartaglia's contributions to the infant science of ballistics were considerable. Equally pivotal was the stand he took as the first man of science to wrestle with the moral implications of his work. After amassing a body of knowledge about the use of gunpowder weapons, he was suddenly overtaken by a sinking feeling. Could a man who understood the naked viciousness of war decently apply his talent to improving a means of mass slaughter?

"It seemed to me that it was a thing blameworthy, shameful and barbarous," Tartaglia wrote. "Worthy of severe punishment before God and man, to wish to bring to perfection an art damageable to one's neighbour and destructive to the human race and especially to Christian men in the continual wars they wage on one another."

In a fit of remorse, he destroyed all his notes and writings on the subject of ballistics. Teachings on these matters, he felt, were a "shipwreck of the soul." His impulse was a modern one, driven not by a superstitious identification of gunpowder with the devil, but by humanistic anguish over the tribulations of war. His childhood memories and broken speech never let him lose sight of the nightmare that weapons visited on helpless victims.

Soon afterward, though, the French king forged an alliance with the Ottoman sultan to plan a joint invasion of Italy. Tartaglia again felt contrite, this time for having neglected studies that could help Christian artillerymen fight off "the ferocious wolf preparing to set on our flock." He resolved to reconstruct his ballistic findings and convey them to the military authorities as quickly as possible. In this way, he enacted an ethical struggle between necessity and compunction that would distress scientists down the ages. In the haunted eyes of atomic-bomb researcher Robert Oppenheimer we see a reflection of Tartaglia's dilemma four hundred years earlier.

———

AT THE END of the sixteenth century, ballistic questions that had confounded Tartaglia were taken up by the towering genius Galileo Galilei. Born in Pisa in 1564, Galileo shared with Tartaglia a difficult financial situation and a stint as a teacher of mathematics. When he addressed the motion of projectiles, Galileo authoritatively wiped away the erroneous ideas left over from classical theories, particularly those of Aristotle. Most astoundingly, he held that impetus was not, as earlier theories had proposed, a reservoir of motion that, like heat, gradually dissipated. Contrary to intuition, it was an inexhaustible force that could only be counteracted by another force.

Galileo took four basic steps in his effort to understand the trajectory of a cannonball. First, he imagined that the ball encountered no air resistance. This enabled him to proceed on a purely mathematical basis, and to radically simplify the problem. Second, he resolved the movement into its parts: the motion imparted by the gunpowder versus the movement resulting from the pull of gravity, a misunderstood force still identified with the "natural" motion of an object. Both forces acted on the ball continuously. Third, he proposed the concept of inertia. A body at rest remained at rest, a body in motion continued in motion—only an outside force could alter either state. Finally, he defined acceleration as the change of velocity over time resulting from the application of an outside agency. Where Aristotle had held that a force needed to be applied to an object to keep it moving, Galileo stated that force was needed only to accelerate a projectile, that is, to change its velocity. Nineteen centuries of assumptions about mechanics came crashing down.

If the force of the exploding powder and the force of gravity acted equally on the projectile, its path would be a straight line angled downward. This was not the case, Galileo said. Gravity constantly accelerated the fall of the object, while the powder, with its one push, set it in motion at a constant velocity. Therefore, the real path was a curve, one that tended increasingly in the direction of the earth. Always passionate about geometry, Galileo found the idealization of all trajectories in the parabolic curve. The force of the gunpowder and the angle at which the barrel pointed determined the dimensions of the parabola, but did not alter its shape. Every cannonball traced a predictable curve.

Galileo's theories did not, in fact, mirror reality. Air resistance introduced elements that would over the years require complex mathematics and rigorous experiments to unravel. Yet gunpowder and cannon had provided a focus for the scientific investigation of reality, one that overturned long-standing error and laid the foundation for a rational age. By the end of the seventeenth century Isaac Newton was explaining planetary motion by the example of a cannonball fired horizontally from a mountain with greater and greater force until its trajectory carried the ball into orbit. The movement of heavenly bodies, he asserted, reflected the dynamics of artillery projectiles.

––––

JUST AS THE speculation of the earliest chemists did little to advance the formulation of gunpowder, the theories of mechanics and ballistics contributed little to straight shooting. "At 200 yards with a common musket," a British colonel said as late as 1814, "you may as well fire at the moon." During the 1700s, it was estimated that fewer than one-half of one percent of musket shots hit their targets. A soldier, it was said, had to fire seven times the weight of an enemy in lead in order to kill him. This might have been an exaggeration, but it pointed to a demonstrable truth. In a 1742 battle, Prussian forces fired 260 rounds for every Austrian killed. Lack of accuracy was the overriding trait of gunpowder weapons during most of their history. The unpredictable flight of projectiles influenced tactics on land and sea, from massed musketry to yardarm naval battles. During the English Civil War of the 1600s a royalist colonel condemned to death beckoned his firing squad to come closer for fear that inaccurate shots would botch the execution.

The man who brought this problem most clearly to light was the Englishman Benjamin Robins. Born in 1707 to Quaker parents, he turned his back on the pacifism for which his family's sect was noted and applied his genius to military affairs. Robins was dissatisfied with the tools available for testing gunpowder. He thought studying actual projectiles was the key to understanding powder's dynamics. He invented the ballistic pendulum, which consisted of a large wooden block suspended from an arm capable of swiveling. The block absorbed the energy of a musket ball and revealed, by the length of its swing, the ball's momentum. Using it, Robins for the first time determined the actual speed of a bullet, which he reported as 1,139 miles per hour.

Modern studies confirm that projectiles from both muskets and cannon of the era did have a high initial velocity, probably between 1,000 and 1,200 mph. This is only half the speed of a modern rifle bullet, but considerably faster than the 750 mph velocity of sound.

Robins discovered a related fact that astounded weapons experts. Projectiles were drastically affected by their struggle to move through air. The effect of drag, he determined, was 85 times as great as the influence of gravity. A sphere was the opposite of aerodynamic. As a ball forced a path through the thick soup of air, it encountered much more resistance than a cone or oval projectile of the same weight. Round musket balls generated four times the drag of modern bullets. As a result, they lost half their speed during the first 100 yards of flight. It didn't take much more slowing before they were no longer lethal on striking a man. A cannonball was subject to forces equivalent to those retarding the musket projectile, but because of its greater mass, drag did not affect it so quickly.

In the investigations that he summarized in his 1742 book *New Principles of Gunnery*, Robins verified facts about gunpowder that were, if anything, more disturbing than the information about air resistance. He began the scientific inquiry into the reason for the persistent, almost ludicrous inaccuracy of gunpowder weapons. To do so, he clamped a musket in a rest and measured its performance by firing through paper screens set 50, 100, and 300 feet in front. By the time the ball reached the middle screen, it was off a straight line of flight by 15 inches. At 300 feet, the deflection was almost six feet.

Tests like Robins' said nothing about the marksmanship of soldiers and everything about a flaw built into the musket itself. The reason for the divergence from the point of aim was one familiar to any golfer who has ever sliced a ball into the woods, any tennis player who has faced a wicked topspin. Spin deflected a moving sphere from its original course. Rapid rotation created a variation in air pressure on either side of the ball, pushing it away from a straight path. As a result, a musket fired at a man from 100 yards away had only a fifty-fifty chance of hitting him.

Why did the musket ball spin in the first place? The sphere was made to fit somewhat loosely inside the gun. The gap between projectile and bore, known as "windage," eased loading and provided a measure of safety if breech pressure mounted too high. With the push of the exploding powder, the ball glanced off the inside of the barrel here and there as it traveled toward the muzzle. The last point of contact determined the speed and direction of its rotation. Since both were unpredictable, the ball's path from one shot to the next could vary wildly, veering right or left, lifting or sinking.

Robins illustrated the effect by bending the barrel of a musket four inches to the left before he fired it through the screens. The ball started out as expected, veering sharply to the left. But by the time it reached the last screen it had actually swerved back to the right of center. The leftward bend had forced the ball to scrape along the right side of the bore, imparting a clockwise spin that resulted in a severe slice to the right.

The inherent inaccuracy of musket fire made attempts at marksmanship almost worthless. Muskets lacked sights. In the British army the command that preceded "fire" was "level" not "aim." Soldiers were not expected to pick out individual targets, only to remain in synch with their fellows so that volleys exploded in unison.

Cannon too, Robins found, could be seriously inaccurate. At a range of 800 yards, the ball from a field gun strayed from its target by as much as 100 yards, and successive firings under identical conditions might fall 200 yards apart. The ball did continue to carry a lethal punch—a 24-pound ball still traveled at supersonic speed after

600 yards. But even experienced gunners had to rely on luck to place a 4-inch ball on a target half a mile away.

The sphere was clearly not the ideal shape for ammunition. It was used because it allowed for quick loading and a rapid rate of fire. What's more, an elongated projectile, though preferable, would have been subject to tumbling, further distorting its path. What was needed, Robins knew, was to impart to the bullet a predictable spin at right angles to its path. This spin would act like a gyroscope, counteracting the projectile's tendency to swerve or tumble. It was the principle of the rifle.

Gunsmiths had been familiar with rifling for more than two centuries. Its discoverer was unknown, but his invention would prove one of the most important in the history of firearms. The idea was simple: A series of grooves were scored along the inside of the gun's bore, gently twisting from one end to the other. As the ball scraped along them, the grooves imparted a rapid spin. The Turin armory had a rifled gun as early as 1476. Rifles of high quality were available across Europe, especially in Germany, by the first quarter of the 1500s.

Certainly the benefits astonished the first shooters of rifles. As if by magic their firearms had become much more accurate than smoothbores. Searching for an explanation, magic was what they hit upon. In 1522, a Bavarian necromancer who went by the name Moretius provided a clear explanation of rifling's effect. The paths of ordinary bullets, he declared, were influenced in flight by spirits, the little imps with whom frustrated shooters were familiar. The shot from a rifled gun flew straight because no demon could remain astride a spinning ball. As evidence he cited the spinning heavens, which were free of demons, and the motionless earth, crowded with them.

Like many theories involving the supernatural, Moretius's idea provoked controversy. Other metaphysicians put forth the equally plausible notion that gremlins could only ride on spinning orbs, that the balls from rifled guns were demon-guided toward their targets. In March of 1547 the Sharpshooters' Guild of Mainz in central Germany put the matter to the test. They fired twenty ordinary lead balls from rifles at targets 200 yards distant. From the same firearms they shot twenty balls molded of pure silver, each triply blessed and marked with a miniature cross. Nineteen of the profane balls found the target, none of the sacred. The issue was decided—demons favored spin. Church officials banned the diabolical rifles in the city, citizens burned them in the town square. It's more likely that the silver, unlike the softer lead, did not adequately grip the gun's rifling or that the inscribed crosses added instability to the holy bullets. In any case, the ban on rifles was soon forgotten by hunters intent on bagging game.

Muzzle-loading rifles, though accurate, were beset by a significant drawback. In order for the grooves to do their work, the ball had to fit between the bottom of one groove and the bottom of the groove opposite so that the ungrooved sections of the barrel, the lands, bit into the lead as it passed along. Rather than being slightly smaller than the bore of the gun, the ball needed to be slightly larger. To load it, the shooter hammered it home, inch by inch, using an iron ramrod and a mallet. For hunters, the laborious loading was an inconvenience. During a battle, the delay between shots could be fatal. Rifles were judged inappropriate for war.

In the eighteenth century Robins saw beyond this defect. He predicted that the nation that developed effective rifled firearms for its military would gain a clear advantage. His advice was universally ignored. Commanders continued to favor massed volleys of inaccurate musket fire.

In 1751, Robins' expertise in fortifications won him a job with the British East India Company. In Madras he caught a fever and died, at age 44, while in the process of drafting a report on his work. Though he had mapped a path by which gunnery could move away from guesswork and rules of thumb, his work would find little practical application anytime soon.

———

THE FAILURE OF commanders to pay much heed to Robins' startling findings about the inaccuracy of gunpowder weapons is part of a curious stasis that gripped military technology from the end of the Thirty Years War in 1648 until well into the 1800s. In spite of the fact that men of science were gradually gaining an understanding of the rules that governed gunpowder's actions, military leaders did not seek out ways to increase its effectiveness. The development of gunpowder weaponry came to a virtual standstill. For two centuries, troops shot at each other with smoothbore muskets and muzzle-loading artillery.

This element of restraint, this reluctance on the part of governments and commanders to pursue innovation, represented a tacit understanding among the European elite that war had become too brutal and too destructive. Other, more practical factors were certainly involved as well. For one thing, the cost of guns meant that re-equipping soldiers with innovative arms was tremendously expensive. Cash-strapped states were likely to stick with the firearms they had and with tested methods of combat. But the notion of a prolonged period of implicit arms control is intriguing. During this period, convention, formality, etiquette, even a theatrical quality all influenced how wars were fought.

In 1625 the Dutch legal scholar Hugo Grotius published *On the Law of War and Peace*, the first modern attempt to impose codes on the conduct of war. In this seminal work, Grotius argued for the humane treatment of soldiers, citizens, and property. His effort was a first step toward defining an international society of states governed by natural laws. It was sign of a broader willingness to rein in gunpowder's effects. "Forbearance in war is not only a tribute to justice," Grotius wrote, "it is a tribute to the greatness of the soul."

The eighteenth century was hardly a period of general peace. War sputtered through Europe, spread to areas of European conquest around the world, flared into the epic struggles that followed the French Revolution. But commanders and individual soldiers marched to all these wars armed with weapons whose basic form would have been familiar to their grandfathers and great-grandfathers. Solid shot and smoothbore musketry, as fearsome as they appeared to the soldier on the field, did not fully exploit gunpowder's capacity to kill. There were no developments comparable to Charles VIII's shattering artillery of 1494 or to the widespread adoption of infantry firearms during the 1500s. The weaponry available to armies was virtually the same across national borders. Warfare became a chess game for kings.

In this age of formality and ceremony, drill became not just a means of increasing soldiers' firepower but a strict form of discipline that held troops in check. To pursue and slaughter a defeated army was viewed by officers as dishonorable, as was the killing of prisoners, a common practice during the Thirty Years War. Fellow officers in particular were to be treated with courtly respect. Louis XIV of France in 1705 gave his commanders permission to surrender a bastion honorably after one small breach and the repulse of one assault—an officer no longer needed to defend a fort to the last man.

———

RESTRAINT INFLUENCED gunpowder's use away from the battlefield as well. The custom of two men meeting at dawn and firing pistols at each other to settle a quarrel is one of the curious artifacts of gunpowder's history, an emblem of the absurd lengths to which formality was taken during the age. The first American treasury secretary Alexander Hamilton famously met his end in an 1804 duel with his rival, Vice President Aaron Burr. British prime minister William Pitt the Younger and the Duke of Wellington both faced political opponents on the dueling ground.

The gunpowder duel gained in popularity during the eighteenth century and continued well into the next. Like other protocols of the era, the custom hearkened back to the days of chivalry and contests of single combat. Strict conventions made pistol dueling as much a rite as a fight. The seconds worked out the rules of the engagement beforehand. The men stood at marks an agreed distance apart. At a signal they fired simultaneously. Since smoothbore pistols were the rule, accuracy was partly a matter of chance. Having "stood fire" came to be seen as a sign of honor. Sometimes the relieved disputants left the field the best of friends.

If the worst happened, a dueling manual advised that the wounded party "not be alarmed or confused," and if he died to "go off with as good a grace as possible." Or, as the saying at an English tavern famous as site of duels went, "Pistols for two and champagne for one."

In spite of its romantic imagery, the pistol duel was in reality a crude, pseudo-heroic travesty of both justice and honor. In 1792 the American author Hugh Henry Brackenridge aptly satirized the convention in *Modern Chivalry* by having his hero answer a challenge by a British officer this way: "If you want to try your pistols, take some object, a tree or a barn door about my dimensions. If you hit that, send me word, and I shall acknowledge that if I had been in the same place, you might also have hit me."

———

Even full-scale battles could be marked by something of the formality of the dueling ground. In 1745, two armies engaged in a conflict that highlighted the nature of war during this long period of gunpowder stasis. The war was one of those pick-up affairs between England and France, each with a diverse collection of allies. The French commander, Maurice de Saxe himself, had once fought against France. The English, with their Hanoverian and Dutch allies, were led by the Duke of Cumberland, son of King George II, already an experienced military man at 22.

The clash took place in Flanders, whose terrain and location made it a convenient meeting place for armies. Saxe made the first move, setting up a leisurely siege of the fortified city of Tournai. Cumberland gathered an army and marched to relieve the city. The two forces, with about 50,000 men on each side, met on May 11 near the village of Fontenoy.

Louis XV came out from Paris to watch the contest, bringing along his 16-year-old son and a few favored courtiers. Dressed in his gold lace, he was pleased to review his resplendent forces. He enjoyed the camaraderie of camp life, sleeping in a barn and telling bawdy stories around the fire on the eve of the battle. Much was made of the fact that this would be the first time since the Hundred Years War, three centuries earlier, that representatives of the royal families of England and France would both be present on a field of battle. The next morning, the king and his entourage took a position on a hill where they could watch the action. Some of the courtiers climbed trees to get a better view.

A disciplined formation of English infantry marched toward Saxe's defensive position that morning, armed with the musket that was in itself a symbol of the stagnation of gunpowder technology. "Brown Bess" had first been issued to British infantrymen in 1703, a smoothbore firearm smaller than the old Spanish musket. Its principal feature was that it needed no smoldering match to ignite its gunpowder charge. Since firearms had appeared on the battlefield in the 1500s, the need for a live source of fire had been an encumbrance. Wheellocks had shown that there were other options. The new system was the flintlock.

To fire a flintlock, a soldier thumbed back a lever against a strong spring. The lever ended in a clamp that gripped a chunk of flint shaped to something of a point. A small pan holding powder adjacent to the touchhole was covered by a lid, the top of which was fashioned into a steel striking plate. When the soldier pulled the trigger, the spring snapped the flint forward. As it struck the plate it simultaneously forced the lid open and showered sparks onto the exposed powder. Flame flared through the touchhole and set off the charge inside. By eliminating the need for a match, this mechanism made firing simpler. The new firearms began appearing early in the 1600s and by the end of the century the matchlock was obsolete.

Supplementing the flintlock was the bayonet, which had taken on the role of the edge weapons that had dominated fighting in the middle ages. This "cold steel," attached under the musket's muzzle, helped

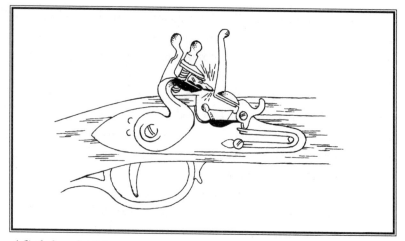

A flintlock musket firing

protect infantry from attacks by cavalrymen and gave them an offensive weapon to use once they fired their volley.

The Brown Bess was designed for speed. Its bore diameter was measurably wider than the width of the .71 caliber bullet, which dropped easily down to the breech. Loading was quick—trained musketeers could fire their weapons every twelve seconds. Facing five volleys a minute from less than a hundred yards turned battle into a test of nerves.

Yet the Brown Bess remained an inaccurate and unwieldy firearm. Key evidence of the stagnation of gunpowder technology is the fact that, with only slight modifications, the gun would remain in use for a remarkable 140 years.

———

SHOULDERING THEIR muskets, the English completed their advance along a ravine and climbed onto the plain of battle as if on parade, drums and fifes sounding, regimental colors aloft. They faced six battalions of French and Swiss infantry. Only sixty feet separated the two masses of troops, the French in bright blue coats, the English in red. The English officers doffed their caps and saluted. The French officers replied in kind. An English captain advanced, a French lieutenant did the same. After toasting the enemy from a pocket flask, the Englishman begged the French to fire first. The Frenchman politely refused the honor. Watching these dueling-ground pleasantries, the courtly spectators on the hill must have reached a pitch of excitement.

Waiting is an awful thing under such circumstances. Some French soldiers grew impatient and let off a few shots at the inviting target. The scattered firing did not move the English line. They stepped closer to the wall of Frenchmen. From thirty feet they extended their muskets and let rip a lethal volley. The French officer who had a minute earlier exchanged courtesies with his counterpart fell dead. Fifty officers and 760 men were chopped down by the fusillade. The French formation wavered, panicked, and ran. "Our affairs began to go badly" was how the official French account put it.

Saxe was not dismayed. The English had won the center of the field, but were cut off from the rest of their forces. He rallied his troops and sent in eight successive waves of cavalry. Finally it was time for the type of massive infantry charge he loved. With the English forces isolated from support, the official account recorded, "The battle was won in ten minutes."

With the fighting over, civility immediately returned. French surgeons afforded the English wounded the same treatment they provided their own men. Prisoners were released on their word of honor. The dead were buried with ceremony. Cumberland sent a note of thanks for these niceties to Louis, signing himself "your affectionate friend."

The bonhomie could not disguise the butchery. More than 5,000 men lay dead, 10,000 wounded. In a human gesture, a suddenly sober Louis took his teenage son on a tour of the Fontenoy field to view the corpses, many of them grotesquely mangled. "Do you see," he asked the dauphin, "what victory costs?"

10

HISTORY OUT OF CONTROL

IN 1774 LIEUTENANT General Thomas Gage was the most powerful man in America, and gunpowder was the focus of his urgent attention. Faced with radicals fomenting rebellion, the supreme military commander in North America and Royal Governor of Massachusetts followed the instincts of a sea captain beset by murmurs of mutiny— first, secure your powder. Gunpowder had become the principal means of making war and the volatile fuel of social unrest.

Since the colonies had almost no gunpowder mills, control of this strategic commodity seemed a sure guarantee of peace. A good portion of the supply in Massachusetts was stored in the Provincial Powder House built on a remote hill six miles north of Boston. To seize the powder, Gage needed to act quickly. If rebels got wind of his plan they could very well oppose the move or spirit the powder away in advance.

Gage's experiences with gunpowder violence had been bitter ones. In 1745 he had served in the British force defeated at Fontenoy. His luck did not improve when, assigned to America, he accompanied the blunt and imperious General Edward Braddock on his 1755 campaign to drive the French from the Ohio Valley. Set upon by a force of Indians and Canadian militia, the redcoats were decimated. Gage's rear guard allowed the escape of survivors, including a provincial officer named George Washington.

Conservative, devoted to the rule of law, the hard-working Gage was out of his element dealing with the quick-witted radicals who were egging on the Boston rabble. "Too honest," one observer called him, "to deal with men who from their cradles had been educated in the wily arts of chicane."

Honest or lacking imagination, Gage in 1770 had sent the notorious 29th Regiment of Foot into Boston to quell a disorder, leading to the spasm of musket fire and five deaths that the colonials called a massacre. He had recommended the punishing Coercive Acts in response to the dumping of tea into Boston Harbor in 1773, closing the port and curtailing the town meetings that bred the vice of "democracy." Yet Gage remained cautious. Married to an American heiress and keenly aware of the risks of open conflict, he sought above all to avoid war.

At 4:30 on the morning of September 1, 1774, Gage sent a company of soldiers rowing across Boston Harbor in longboats. They marched to the powder house, a windowless stone tower, and removed 250 barrels of gunpowder, along with two brass field guns, securing them in the main British stronghold on Castle Island. The reaction to this operation confirmed for Gage the wisdom of his caution. Rumors tore through the land: Boston had been bombarded, six were dead, war was imminent. Patriots lit beacon fires and tolled church bells for hours. By the following day the country was teeming with armed patriots. Twenty thousand men from the Connecticut Valley alone took to the roads. Whig leaders struggled to restrain enraged citizens. Notorious Tories fled for their lives. It was a frenzy that would become known in New England as the Powder Alarm.

Gage abandoned as too provocative a plan to send troops forty miles inland to confiscate the store of powder from Worcester. Instead, he forced all Boston merchants to sell their stocks of gunpowder to the Crown. He set up cannon and fortifications on Roxbury neck, which connected Boston to the mainland. He urged his London superiors to assign him 20,000 more men. Considering that Britain's peacetime army consisted of only 12,000 infantrymen, the request gave a hint of Gage's uneasiness. The high command sent him 400 marines.

For their part, the rebels established a committee of "mechanics" to keep an eye on British movements with the hope of forestalling any more powder raids. The thirty Boston volunteers included the silversmith Paul Revere. "The spirit of Liberty was never higher than at present," the 40-year-old Revere had written on September 4. "The troops have the horrors."

In October, King George himself issued an order banning the import of gunpowder to America and decreeing that all supplies be secured for the Crown. Again the alarm went out. That December, Paul Revere rode fifty miles in a raging snowstorm to Portsmouth, New Hampshire, to report that the British regulars were coming there to confiscate the gunpowder at Fort William and Mary.

Before the British reserves arrived, four hundred local militiamen marched on the fort. The outnumbered garrison managed to fire three cannon shots, hitting no one, before being overwhelmed. The rebels had the effrontery to haul down the king's colors before breaking into the magazine and removing the powder. One hundred barrels of explosive were hauled out by cart and boat.

Gage watched events reeling out of control. At Newport, Providence, and New London, rebels removed gunpowder from depots and hauled it to the safety of the interior. In February 1775, Gage heard that ships' cannon were being converted to field pieces in Salem. The force he sent to investigate found themselves eyeball to eyeball with a cadre of Salem militia and Marblehead fishermen. The troops backed down. "Things now every day begin to grow more and more serious," wrote Hugh Percy, Gage's loyal subordinate.

The king himself was livid over the theft of his powder and desecration of his fort. London urged action against the patriots, "a rude Rabble without plan, without concert, and without conduct." The British commander was still convinced that the control of gunpowder was the key to defusing the threat. But any further operation to put the volatile substance under lock had to succeed—another fiasco like Portsmouth or Salem would be disastrous.

For his target, Gage selected Concord, fifteen miles northwest of Boston. The town was a center of rebel sentiment and supplies. One building, his spies reported, held seven tons of gunpowder. Gage had to strike at the rebels before they could muster their forces, which vastly outnumbered the troops under his command. The secret operation began at 10 P.M. on the night of April 18. British troops in Boston were awakened with a whisper and told to bring 36 cartridges of powder and ball. They slipped out the back doors of their barracks and made their way through the quiet streets. A dog barked—a soldier silenced the animal with his bayonet. Sailors ferried the 900 men across the Charles River. At two in the morning, wet and shaking with cold, the regulars began their march.

They had not been quiet enough. They began to hear, off in the distance, the ominous sound of clanging church bells and the crack of warning shots. Revere and his fellows had caught wind of the movement, crossed the river before the British, and were even now spreading the alarm through the countryside.

Gage had put two veterans in charge of the raid: the corpulent and careful Lieutenant Colonel Francis Smith, and the experienced Marine Major John Pitcairn. Pitcairn had a professional's disdain for the noisy rabble. "If I draw my sword but half out of my scabbard, the whole banditti of Massachusetts will run away," he wrote in a letter.

"I am satisfied they will never attack regular troops." Sensing the unrest spreading through the countryside, Smith sent Pitcairn ahead with six companies of light infantry. At 4:30, with the first gray light seeping into the sky, Pitcairn ordered his men to load their muskets. They bit off cartridges, poured in powder, rammed the balls home. They were about to march into the village of Lexington.

When they reached the triangular common in the center of the hamlet, they came face to face with a small group of the Middlesex County militiamen. The redcoats broke out of their column and formed a line. Bystanders poured from nearby Buckman tavern. Other townspeople watched from the surrounding roadways, only vaguely aware that they were witnessing an event of historic moment.

The two sides eyed each other across sixty yards of patchy grass. The soldiers were unlettered, hard-drinking men far from home, low in spirits, despised by the colonials around them. They likewise scorned the men opposite as "rebels," "provincials," "Yankeys." Those militiamen, long hair tied back in queues, were hardly dreamy idealists. Many had lived through bloody encounters with Indians and French troops in the wilderness. Yet in spite of the mutual contempt, those on both sides of the green were, up to that moment, fellow countrymen, members of the British nation.

Some militiamen thought it foolhardy to stand in the way of the regulars. Their commander, Captain John Parker, silenced debate. "Stand your ground. Don't fire unless fired upon. But if they want to have a war let it begin here." His men stood. The situation grew tense. The British soldiers began their distinctive battle chant, a menacing "Huzzah! huzzah! huzzah!" British officers screamed at the rebels to lay down their arms. Parker had second thoughts. With confusion mounting, he ordered the militia to disperse. Some did, "though not so speedily as they might have done," one witness reported. Some stayed where they were.

Who fired the first shot that morning will never be known. The apprehension and animosity that had been growing for ten years suddenly came to a head. A shot, maybe two, rang out. The British infantry, though renowned for their discipline, began firing sporadically without orders. Some of the militia returned fire. Then came the awful, ripping sound of a full volley of musketry. Almost instantly, one witness reported, "the smoke prevented our seeing anything but the heads of some of their horses."

Chaos descended on the green. The British regulars began to reload and fire with the speed conveyed by years of relentless drill. Horses bolted. Men ran and were shot down. Spectators fled in panic. One patriot, chased into the village meeting house, which served as an armory, thrust his firearm into a barrel of gunpowder, ready to blow the building up if the redcoats pursued. Jonathan Harrington, who lived in the village, fell with a ghastly wound to his chest. He crawled toward his home and died on his doorstep in front of his wife and son.

British officers ordered the drummers to beat the men to arms. Reacting automatically to the signal, the troops reassembled. They marched on, leaving eight dead militiamen behind them. They found no gunpowder in Concord—Revere's warnings had allowed the patriots time to haul most munitions out of the village. After knocking the trunnions off several cannon and cutting down the town's liberty pole, the regulars formed up for the long march back to Boston.

The clash that took place along that route has been enshrined in American mythology as an example of individualistic Indian fighting. Longfellow described "How the farmers gave them ball for ball/From behind each fence and farmyard wall." Isolated skirmishers and ambushes played a role in the action, but most of the battle involved attempts by the American militia to stand up to the redcoats in conventional formations.

The events of the day were far worse than anything General Gage could have imagined in all his anxiety over gunpowder. His best troops had been smartly manhandled by a band of determined farmers and merchants. "The Rebels," he wrote, "are not the despicable rabble too many have supposed them to be."

Gage's bad luck continued. He found himself besieged in Boston. By August, his own officers were refusing to obey him. He was recalled to England in October. By that time, a full-scale war had begun in North America.

———

IN JUNE, Gage assigned the duty of making the first serious strike against the rebels to the man who would replace him, General William Howe. The Americans had suddenly fortified Breed's Hill on the Charlestown peninsula across the harbor from Boston. Military custom, an inflated sense of honor, and more than a trace of overconfidence all dictated a frontal assault on the ragtag rebel troops.

Dressed in their red woolen coats on a warm, sky-blue spring day, the British infantrymen struggled up the slope. One of the rebel commanders, Colonel William Prescott, was painfully aware that his men lacked gunpowder for more than a few volleys. His advice, "Don't fire till you see the whites of their eyes," became an adage for the history books, though similar words had been attributed to a Scottish lieutenant colonel in 1743. The sentiment was pure economy. The Americans waited, holding off their volley until the British regulars came within ten yards of their hastily built fortifications. The sudden crash of fire directly into their faces sent the British regulars reeling. They retreated to the bottom of the hill.

Howe and his officers, shaped by the conventions of European warfare, felt that to call off the assault or switch their direction of attack would be a stain on British honor. They ordered their men back up the steep hill. Again, disciplined American fire repulsed them. Nothing would do but a third frontal charge.

This time, no volley met the redcoats' advance. The Americans had run out of gunpowder. Lacking bayonets, they were overrun and forced to retreat. The battle had been costly for the British. Of the 2,200 soldiers who had advanced, half had been killed or wounded.

"A dear bought victory," a British officer noted, "another such would have ruined us."

The Americans' failure to hold their position in the battle named for nearby Bunker Hill was a result of tactical ineptness and logistical oversight. But the incident was a reminder of a critical weakness in the rebels' plan to take a stand against the King's troops: They desperately lacked gunpowder.

Powder making was not unknown in America. During the French and Indian War of the 1750s, entrepreneurs had established a few small mills. But when peace returned, the government in London discouraged the trade, along with colonial manufacture in general. Royal governors instead imposed a tax on ships putting in at American harbors and earmarked the money to buy gunpowder made in England. In any case, the local product could not have competed with the powder ground in large English mills using the best Indian saltpeter.

As the nascent continental army settled in to a siege of British forces in Boston, the whole enterprise rested on a shaky foundation. In August 1775 George Washington wrote from Cambridge, "Our situation in the article of powder is much more alarming than I had the most distant idea of. We have but 32 barrels."

This was enough to issue each man only about half a pound of powder. By the end of the month the supply had dwindled further. Firing powder-hungry artillery was almost out of the question. Looking on Boston from Prospect Hill, General Nathaniel Greene lamented, "Oh, that we had plenty of powder; I would then hope to see something done here for the honour of America."

In the early stages of the war, the American forces had to beg, borrow, and steal gunpowder. An inventory of all thirteen colonies found only about 40 tons on hand, enough for a few months of operations. About half was sent to Cambridge to supply Washington's army, the rest allocated for local defense. In June 1775 a hundred pounds of gunpowder could not be purchased in New York City at any price. A band of "Liberty Boys" in Savannah, Georgia, made off with a few barrels of powder from the government magazine in May, and returned in July to capture fully six tons of the precious explosive from a ship in the harbor.

The Second Continental Congress discussed the gunpowder shortage as one of its first orders of business, establishing New York and Philadelphia as depots for the collection of saltpeter and sulfur. They decided to offer subsidies to manufacturers and to recruit French experts to teach Americans the fine points of powder making. The possibility was raised of arming rebel troops with pikes and staves instead of gunpowder weapons. Benjamin Franklin offered well-reasoned arguments for adopting the bow and arrow in place of the musket, but by the eighteenth century a return to the fighting tactics of Henry V at Agincourt was no longer a viable option.

By Christmas of 1775 Washington stated flatly, "Our want of powder is inconceivable. A daily waste and no supply administers a gloomy prospect." In the middle of January, the supply was virtually gone. By contrast, General Howe possessed tons of powder and could rely on a steady supply arriving by ship. If he had decided to march out of Boston, he could have crushed Washington's small army and ended the rebellion. He waited too long.

A year earlier, an article in the *Royal American Magazine,* illustrated by Paul Revere, had discussed methods of producing saltpeter at home. The essay pointed out that saltpeter is "an effluvia of animal bodies. Pigeon houses, stables, and barns but especially old walls are full of it." Since the encouragement of preparations to attack His Majesty's forces could be labeled seditious, the writer extolled the medicinal uses of the salt and recommended it as a flavoring for brandy. Readers got the point.

Once hostilities had begun, committees of safety and other revolutionary groups published numerous pamphlets and broadsides encouraging a cottage industry in saltpeter making, offering detailed instructions. Saltpeter, though, was useless without facilities to convert it into gunpowder. The provincial congress in Massachusetts decided to set up a mill in Canton, twelve miles south of Boston. For guidance, they sent the intrepid Paul Revere to learn the details of powder making in Philadelphia, where "the manufacturing of powder is carried on with Considerable dispatch and advantage." The most prominent operation, in nearby Frankford, was run by Oswell Eve, an aging ship captain and merchant. He hurried Revere through his works, refusing to allow him to talk to workmen. Eve was later exposed as a Tory, his mill confiscated.

Much of the locally produced powder was of decidedly inferior quality. Massachusetts general William Heath complained that the powder his men received was "bad." Washington suspected "there must be roguery or gross ignorance in your powder-makers." In the eighteenth century, European gunpowder makers, drawing on centuries of tradition and craft secrets, continued to struggle to make a consistent, high-quality, and durable product. Colonial tyros could hardly be expected to master the art overnight.

Local producers did at times make significant contributions to the war effort. Mary and John Patton had set up a small powdermill in the wilds of what would become eastern Tennessee. With her husband serving in the militia, Mary continued to produce the explosive, refining saltpeter in an iron vat and grinding the powder in a crude stamp mill. She supplied 500 pounds of powder to the volunteers who defeated a force of Tories at the Battle of King's Mountain, an important check to British fortunes in the South.

The best hope for easing the American powder crisis lay in imports. During the first two years of the war, fully 90 percent of the new country's gunpowder was either purchased abroad or made from imported saltpeter. The main sources of this crucial supply were French and Dutch traders in the West Indies. A Bordeaux firm sent 2,800 barrels of powder to Martinique and netted a neat profit by exchanging it for tobacco and rum. Congress encouraged the trade by guaranteeing a 100 percent markup for importers. The islands became an active emporium of powder and arms.

An eighteenth-century stamp mill for making gunpowder

Yet a gunpowder shortage remained a worry for American forces throughout the war. In spite of the revolutionary zeal and advice of French experts, shortages of materials, especially saltpeter, continued to hamper production. Even in 1781, before the decisive battle at Yorktown, Virginia, the American supply was judged to be in a "wretched and palsied state." Only determination and a careful marshaling of available powder allowed the Americans to prevail.

———

THE GUNPOWDER that helped the Americans win their independence might have been much more difficult to come by if events in Europe had taken a slightly different turn. The availability of powder for the Revolution was largely due to an extraordinary program initiated by the French government in response to its own shortage of the precious explosive.

In 1774, the French king Louis XV, who had defined the *ancien régime* during nearly sixty years on the throne, died of smallpox. His 20-year-old grandson, on becoming Louis XVI, was alarmed to find that France's self-sufficiency in gunpowder was an illusion. The nation's supply of what had become the military mainstay of all kingdoms was dangerously precarious.

As far back as the fifteenth century, French gunpowder makers, like those of other European kingdoms, had relied on saltpetermen to collect nitrous soil from barnyards and leach the essential salt from the walls of demolished buildings. The king gave these workers the right to confiscate the soil and refuse wherever it was found, a privilege known as *droit de fouille* or right to dig. The perquisites that went with the position were passed on in families and the *salpêtriers* formed a close fraternity.

Country people did not appreciate having their barnyards, outbuildings, and sometimes even their homes dug up. A well-placed bribe could often do the trick—the wealthy were rarely bothered. Sometimes an entire town chipped in to pay the petermen to leave them alone. These bribes became a welcome supplement to the gatherers' income. The entire system was complicated by a byzantine bureaucracy rife with corruption.

By the time Louis XVI ascended the throne, this system had grown hopelessly archaic. England and Holland had been importing cheap saltpeter from India for a century. In Prussia and Sweden, military authorities oversaw efficient artificial saltpeter works. France was among the last of the major powers to rely on itinerant saltpeter gatherers. The *salpêtriers* brought in only half of the 3 million pounds of the material needed every year—the rest was purchased from Dutch sources at steep prices.

Through his ministers, Louis XVI established a governmental Gunpowder Administration, appointing as its head Antoine-Laurent

Lavoisier. It was an inspired choice. One of the most able administrators of his time, Lavoisier was also an exceptionally gifted chemist and ardent patriot. The son of a bourgeois Parisian family, Lavoisier had taken a degree in law. He became wealthy from investments in a private company set up to collect taxes for the crown. The exciting developments in natural philosophy that were sweeping Europe had inspired him to pursue experimental science in his spare time.

To get the gunpowder project rolling, Lavoisier arranged for a contest to elicit the best scientific ideas for increasing the saltpeter supply. But the problem was too pressing to wait for the contest deadline two years off. Lavoisier, always a feverish worker, plunged into a multifaceted program to relieve the shortage. To produce more potassium nitrate from the resources at hand, he encouraged saltpetermen to add ashes or potash to their pungent stew before they extracted the salt. This step, which encouraged the formation of potassium nitrate, had long been a part of niter refining. But the processors needed exact instruction about which type of ashes were best and how much to add.

The greatest impact on supply flowed from Lavoisier's rationalization of the saltpeter industry. He installed proper management, discarded outmoded procedures, and improved record-keeping. A sensible price structure and productivity bonuses encouraged saltpeter makers to invest in better facilities. He reformed, then eliminated the *fouille,* freeing property owners from the hated intrusion of the petermen.

His efforts yielded rapid results. Among the first beneficiaries were the struggling American rebels. Barely a year after Lavoisier took over the Gunpowder Administration, France had enough gunpowder to supply the American war effort. Lavoisier was proud of this achievement: "It can truthfully be said that it is to those supplies that North America owes its freedom," he declared.

France's domestic saltpeter production, only 1.7 million pounds in 1775, had reached 2 million pounds by 1777 and almost doubled by 1788. At that time warehouses held 5 million pounds of gunpowder. Lavoisier had also improved the processes by which powder was

Antoine Laurent Lavoisier (1743–1794)

ground at mills like Essonne, outside of Paris, by adjusting the proportion of ingredients and the time spent grinding the mixture. His country's powder had become the best in the world. The Dutch and Spanish stood in line to buy it. English sea captains complained that they were being outshot by French guns.

The contest that had kicked off the crash niter program turned out to be a fizzle. After a long delay, the prize was finally awarded in 1787 for an idea that involved no scientific breakthrough. Pure science as yet had little to contribute to such mundane problems as extracting saltpeter from sheep dung.

The contest submissions showed that many amateur "scientists" were equipped with more enthusiasm than insight. Some papers relied on a mythical "universal acid" or "vitriolic earth" as the basis of their improvements. One held that saltpeter was a living organism. Another claimed that the author could so accelerate the normal two-year putrefaction process that he could produce saltpeter in three days. One proposal would have required purveyors of beer and wine to conserve their customers' urine in vats.

All of this clumsy theorizing and fantasizing only served to highlight the monumental contributions that Lavoisier himself was making to the advancement of knowledge. His mind was a razor, "the spirit of accountancy raised to genius," one historian called it. His strength was exacting measurement and analysis combined with penetrating insight. Using a balance that could measure to an accuracy of four millionths of an ounce, he probed the transformation of materials by tracking every iota of substance.

The investigation of combustion was never far from the center of his interest. In 1772 he puzzled over the fact that when he burned sulfur the resulting products were heavier than those he had started with. The sulfur was combining with something in the air, he theorized. Five years later he named this substance "oxygen"; it had been first isolated by the English scientist Joseph Priestley. Lavoisier attacked the hundred-year-old theory of phlogiston, which held that fire was a substance hidden inside combustible material. Wrong, he said. Combustion was a process, a chemical reaction. Burning material was combining with oxygen. Oxygen was not a principle of combustion but a physical substance, a gas mixed with the air and locked inside saltpeter. It was the link between respiration and a flame, between natural fire and gunpowder.

———

THE AMERICAN Revolution had again proven that gunpowder could fuel violent action, that its potential energy could be converted to political power. The supply of powder inevitably became a focus of attention during every subsequent period of civil turmoil. On July 12, 1789, turmoil was most definitely growing in Paris as the king contemplated calling in troops to quell his subjects. Concern for the city's gunpowder stocks grew. The governor of the Bastille, Bernard de Launay, ordered the supply of powder stored in the nearby arsenal moved to the fortress-prison. He urgently requested reinforcements for his garrison of eighty-two veterans. Royal authorities sent thirty-two Swiss guards.

The bourgeois citizens of Paris were trying to keep the rage of the lower classes under control. On July 13, electors of the Third Estate distributed arms to an ad hoc militia. Some of the weapons were antique halberds and pikes, but they also included 30,000 muskets and a silver-inlaid cannon that the King of Siam had sent to Louis XIV. On the morning of July 14 hundreds of armed citizens gathered near the Bastille. They had come to seize the fort's gunpowder.

A tense standoff ensued. In the noonday heat, shots were fired. A sputtering battle continued during the afternoon. Balls blasted from the ornamental cannon had little effect on the eight-foot-thick walls of the fortress. Inside, a desperate Launay toyed with the idea of lighting the powder rather than give it up. Cooler heads among the beleaguered veterans dissuaded him.

Lacking both supplies of food and a water source, the Bastille was not prepared for a siege. As evening approached, Launay chose to surrender. He ordered the drawbridge let down. Citizens stormed inside. They freed the seven resident prisoners: four crooks, two madmen, and one political prisoner. Eighty-three Parisians had died in the fighting. The triumphant crowd marched Launay to the Hotel de Ville, where the revolutionaries had set up their headquarters. An irate Launay kicked one of his captors, a pastry cook, in the groin. The citizens turned on him with knives and pistols. The cook sliced through Launay's neck with a pocket knife. The severed head was mounted on a pike. The show had begun.

Lavoisier initially welcomed the Revolution. A new and rational form of constitutional government suddenly seemed possible. But now doubts clouded his outlook. "Those moderate persons who have kept their heads in the general effervescence think that circumstances have led us too far," he wrote to Benjamin Franklin in 1790. "It is impolitic to place force in the hands of those who should be obeying it."

Lavoisier continued to serve his country, to oversee the efficient manufacture of gunpowder, to push forward the frontiers of chemistry. But events were taking on a momentum of their own and rationality was becoming a victim of popular enthusiasm. "Suddenly history was racing out of control," he wrote.

The Revolution continued to accelerate. Lavoisier's weakness for wealth and career as a tax collector came back to haunt him. In November 1793, with the nation in the grip of the Terror, he was arrested for misappropriating government funds. "This affair will probably save me the inconvenience of old age," he wrote as he awaited execution.

Two years after Lavoisier's beheading, the Terror exhausted, the French nation began to shower his name with accolades. By revitalizing gunpowder production, he had made an enormous contribution to the military preparedness of the French Republic, which had had to face grave threats to its existence.

In 1793 the monarchies of Europe, including Spain, Britain, and Russia, formed an alliance to extinguish the revolutionary fire. As the need for powder grew critical, the French people responded. The Committee for Public Safety divided the nation into eight large districts for saltpeter production. The entire population was encouraged to contribute. The nation's leading chemists, as well as simple apothecaries, fanned out through the countryside to offer instructions. Building on Lavoisier's methods, the government drew 6,000 new saltpeter makers into the effort. These were the *salpêtriers sans-culottes*.

Few expressions of revolutionary ardor were as touching as this communal effort. All over the country, amateurs dug up barnyards in search of the vital ingredient for the nation's defense. The success of the program was astounding. In 1794 the revolutionaries quadrupled the best saltpeter output of the Gunpowder Administration. A huge new refinery was established to convert the bounty to a useable chemical. An equally colossal powder factory was set up, quite imprudently, in the heart of Paris. Though this mill soon exploded, the energetic gunpowder program allowed the nation to survive its crisis.

11

THE MEETING
OF HEAVEN AND EARTH

ON NEW YEAR'S DAY, 1800, the *American Eagle* docked at Newport, Rhode Island, sails torn, hull leaking, and food supply exhausted. The ship had taken ninety-one days to cross the Atlantic, a month longer than Columbus had spent making the trip three centuries earlier. One of the passengers who had eaten boiled rats on this godforsaken journey was the man destined to establish the most colossal gunpowder-making operation in the world: Eleuthère Irénée du Pont. With his father, brother, wife, children, and other family members, he gratefully set foot on terra firma this first day of the new century. According to family legend, the troupe celebrated by breaking into a house whose owners were at church and wolfing down a waiting feast.

The family's self-imposed exile had been the brainchild of Irénée's father, the irrepressible Pierre Samuel du Pont de Nemours. An impecunious aspirant to the ranks of the French nobility, Pierre was an idealist and a man of wide connections. He had supported the French Revolution in its early days, had even held the post of President of the National Assembly. He had been luckier than his friend Antoine Lavoisier during the Terror—a delay in the arrival of Du Pont's death order let him escape the guillotine. As the 1790s wore on and the climate for moderates in France failed to brighten, Pierre decided to decamp to America.

He intended to carve a new colony, "Pontiana," out of the woods of Kentucky. Part land speculation, part utopian fantasy, his dream began to deflate as soon as the family landed in America. Thomas Jefferson, whom Pierre had known since the American's days as envoy in Paris, advised him that like-minded plungers had already driven the price of western land to unrealistic heights. The ever optimistic Pierre—he had insisted that the shipboard rats were not that bad a dish—began to search for other opportunities.

While Pierre mulled possible schemes, including a notion of smuggling gold to Spain, Irénée gravitated toward his chosen profession. In France he had learned the gunpowder business under Lavoisier himself. Well-established lore relates that he joined a hunting expedition and was dismayed at the poor quality of American gunpowder. He probably needed no such epiphany to spot the opportunity presented by the weak competition and the growing market among pioneers and the American military. Conditions boded well for an efficient operation run by someone versed in the latest French methods of manufacture.

Pierre had not initially cheered his son's decision to take up the craft of powdermaking—the dirty and dangerous trade hardly suited a gentleman. But as he reviewed Irénée's meticulous estimates of the profit to be made, he swung behind the scheme with typical enthusiasm. Irénée and his brother Victor were soon sailing back to France to raise capital and buy machinery, and Pierre was bragging that the plan "gives us not hope but a positive certainty of great profit." He wrote to Jefferson that Du Pont gunpowder would "send bullets a fifth farther than English or Dutch bullets travel."

Napoleon's government, ever eager to thwart the British, offered the best machinery on reasonable terms. After a fruitless attempt to purchase the established Frankford gunpowder mill outside Philadelphia, Irénée chose a site on the Brandywine Creek near Wilmington, Delaware, and began to build a mill from scratch. The availability of water for both power and transportation influenced his choice, as did the colony of Frenchmen in the area, exiles from a slave revolt in Santo Domingo. Begun in 1802, the dozen buildings that made up the factory took two years to build. In 1804 the plant, known as Eleutherian Mills, shipped its first powder. The Du Ponts sold a total of 22 tons for the year.

Irénée had entered the business at a propitious time. War in Europe blocked competition from that direction even as it increased demand. The U.S. Navy bought 11 tons of Du Pont powder soon after the mill opened in order to subdue the pirates of the Barbary States in North Africa. John Jacob Astor, who was making a fortune in the fur trade, purchased two-and-a-half tons of explosive in 1810. Two years later, as hostilities between the United States and Britain broke into war, yearly sales surged from 25 to 100 tons, then to 250 tons the following year.

In spite of the company's success, Irénée was already mentioning his feelings of "habitual dullness and melancholy," a surfacing of his lifelong tendency toward emotional depression. The nervous strain of days spent dealing with explosives understandably dampened any tendency toward ebullience. Irénée was a hard-working, careful, and exceedingly conscientious man who loathed "all that savors of self-praise and bragging."

Like any prudent powderman, Irénée built his works in anticipation of disaster. Each step in the manufacturing process had its own building. The powder was ground in a series of small water-driven mills constructed of heavy stone on three sides, with a flimsy fourth side facing the Brandywine. This architecture had a grimly practical purpose. An explosion would blow out the weak wall and roof of the building, directing the blast away from the other structures. To powdermen, "to go across the creek" meant to die in an accident.

Irénée took advantage of a number of improvements that powdermakers had introduced as the trade gained an industrial footing. The key process remained the "incorporation" of the ingredients, the pulverizing and grinding that brought the saltpeter, sulfur, and charcoal into their intimate relationship. The mortar and pestle was the traditional method of incorporation, and the French continued to use it well into the 1800s. Powdermen looking for a faster, safer, and less labor-intensive method turned to a type of mill that had previously been used for crushing olives. A pair of stone wheels, five or six feet in diameter and eighteen inches across, stood vertically on an eight-foot-wide stone bed. Connected by their axles to a central shaft turned by a water wheel or horses, the stones rolled at ten rotations a minute in a tight circle. The great weight of the wheels—as much as eight tons—crushed the material spread out on the bed. Powdermen moistened the ingredients with distilled water now, not bishops' urine.

Irénée du Pont began his production process using stamp mills, but soon shifted to these wheels or "edge runners." Originally made from marble, later from iron, the wheels produced a larger volume of more homogeneous powder in a shorter time. Though initially costly, the gargantuan wheels saved considerable labor and eventually became emblems of the industry. In America, powdermen shoveled as much as 600 pounds of ingredients at a time under these rollers. The safety-conscious British, who had banned stamp mills altogether in 1772, limited loads to 40 pounds. For fine rifle powder, wheel incorporation took about four hours.

Pressing was another change that contributed to the more powerful gunpowder of the nineteenth century. Irénée learned of the technique on his trip to France before opening his mill. Workers shoveled the incorporated powder, known as "millcake," from beneath the

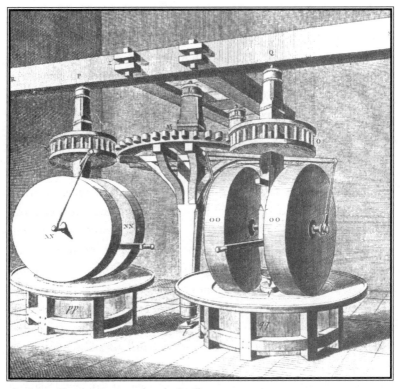

An eighteenth-century gunpowder wheel mill

wheels into a box and applied a force of 1,200 pounds per square inch, using a screw press. The slightly damp powder was squeezed to roughly half its former volume, forming two-foot-square slabs of "presscake" with the hardness of slate. Pressed powder was considerably denser than ordinary corned power—a smaller amount of explosive packed more punch, allowing the American military to load its cannon with one-third less powder.

When Irénée du Pont started his operation, he grained his product by forcing moist millcake through sieves. But pressed powder could not be broken into granules using this method. To crush them, workers had to pound the hard slabs with hammers or feed them through rollers. A series of sifters sorted the granules of powder by size. Standard musket powder was classified as F grade. Finer powder used for rifles was designated FF or 2F. Still finer grades used for pistols and as a primer were classified 3F and 4F. As a last step, the powder was tumbled in glazing barrels. Hours of agitation rounded the edges of the grains, increasing the powder's durability. The finished grains of rifle powder had the consistency of granulated sugar, whereas those of cannon powder were the size of uncooked rice kernels.

One other significant advance increased the force and consistency of nineteenth-century powder. Instead of burning charcoal in earth-covered pits, manufacturers distilled it by heating wood in closed iron retorts. By carefully controlling the temperature, workers could influence the explosive qualities of the powder made from the charcoal. This kiln charcoal produced gunpowder that was both stronger and more consistent than earlier versions.

In spite of the industrialization of these processes, the gunpowder trade retained a considerable element of craft. The powderman had to use his judgment about the amount of moisture to add, how long to grind the materials, the amount of pressing needed. Powder making was a skilled and specialized trade, and experienced workers developed an uncanny ability to tease a consistent product from the ingredients.

Danger remained one of the constants of the trade. Workers were well aware that the material they handled every day could, in an instant, kill them. Precautions were taken: shovels were made of wood, workers wore shoes with wooden pegs rather than hobnails, horses went unshod, men chewed tobacco rather than smoked. The Du Ponts adopted an enlightened policy toward their employees, providing housing and assuring the men that their families would be cared for in case the worst happened.

On March 19, 1818, it did. Residents of Lancaster, Pennsylvania, 43 miles northwest of the Brandywine, felt the ground tremble. In Wilmington itself, doorbells rang spontaneously, dishes broke, houses rocked. The Du Pont works had exploded.

An 1851 advertisement for Du Pont gunpowder

Thirty-six workers were killed instantly. Powdermen's houses a half mile away were crushed. Irénée's beloved wife Sophie was injured inside her own home—she would never fully recover her health. Human remains were found dangling from treetops. The carnage appalled the visiting dignitary Marshal Emmanuel Grouchy, who had braved the field of Waterloo three years earlier. All the Du Pont homes were turned into hospitals. In the aftermath, widows were allowed to stay in company houses on pensions.

This accident was one of many—they hit the mills on average once every 14 months. The damage often obscured the cause, though the solid, ingeniously designed mill buildings usually survived. A bit of a nail or a piece of grit in the mixture could set off a spark during processing. A wagon wheel could jar against a stone. An excess of friction, a moment's carelessness, an unattended candle could mean an instant calamity.

By the 1830s Irénée had cleared away most of the family's debts and put the business onto a profitable footing. He died in 1834, leaving the business to his children. The eldest, Alfred, who had made the harrowing trip across the ocean as a toddler, took over the operation of the mills. He was joined by Alexis, only 18. The middle brother, Henry, was serving in the army.

Alfred faced both problems and opportunities. His father had established a network of agents around the country, but had always centralized production in Wilmington, where he could oversee the details. Transportation presented a difficulty. The railroads that began to crisscross the landscape during the 1840s often balked at carrying gunpowder. Engines spewed cinders, the clanking of machinery generated sparks, and trainmen wanted no explosives on board. The Du Ponts employed Conestoga wagons shepherded by daredevil drivers to make the six-week trip to Pittsburgh. When three of these wagons blew up in downtown Wilmington, laws were passed banning them from the city.

At the same time, the frenzy of canal and railroad building, combined with the opening of the West, brought steady demand for powder. Stumps and rocks needed to be blasted. Pioneers demanded the Du Ponts' quality hunting powder. Miners had benefitted in 1831 when Englishman William Bickford invented a fuse consisting of gunpowder-impregnated twine, making blasting considerably safer. The

The Du Pont gunpowder mills after a serious explosion

Mexican War, which broke out in 1846, prompted the U.S. government to purchase a million pounds of gunpowder. The Du Ponts expanded their mills and ran them around the clock to turn out 5 tons of powder a day.

Cranking up the pace of work increased the danger. On April 14, 1847, the works exploded again. "In an instant, without the slightest warning," wrote a family member, "there came a shock that seemed so terrific in its nature that I could only compare it to the meeting of heaven and earth. It appeared not to be local but a crash of the world." A row of buildings blew up in a quick chain reaction. Stones and beams rocketed into the sky. Glass shattered in nearby homes. Doors burst in. The smell of burnt powder choked the air.

This explosion killed eighteen and injured many others. The accident broke Alfred's nerves. Some of the workers who died, who were quite literally blown to bits, were men he had grown up with—their fathers had worked for his father. His health failing, Alfred retired three years later.

Family members were not spared the calamities of the dangerous enterprise. In 1857 a powder explosion took the life of Alfred's younger brother Alexis at the age of 41. That left Henry as the man in charge of the steadily growing powder-making firm. Enterprising and aggressive by nature, Henry would guide the company to utter domination of the gunpowder market in America.

———

THE SAME YEAR that the Du Pont family was landing in the New World, an English chemist named Edward Howard presented a paper to the Royal Society in London describing the preparation and properties of mercury fulminate. The compound had a quality that distinguished it from almost every other known chemical: It was highly explosive. It didn't even require a match to set it off, a sharp concussion was sufficient.

Fulminates had been known for two hundred years. They were derived from the combination of a metal and an unstable organic acid related to ammonia. They took their name from the Latin word for "lightning." Silver fulminate, so sensitive that a hint of friction was enough to make it erupt, was a staple of practical jokers. An 1818 fireworks manual suggested that a pinch inserted into the end of a cigar held the potential for laughs.

In 1800, Howard had a more momentous purpose in mind for mercury fulminate. He hoped to find a chemical replacement for gunpowder, which had reigned as man's only practical explosive for almost nine hundred years.

He tried the synthetic chemical in a gun, but the barrel burst. Later, a more serious accident destroyed most of his laboratory, injured him severely, and convinced him that he was "more disposed to prosecute other chemical subjects."

This was the heyday of the amateur scientist and Howard's findings did not pass unnoticed. One who continued the work was Reverend Alexander Forsyth, a clergyman outside of Aberdeen, Scotland, and an avid sportsman. Forsyth was bothered by a problem that had accompanied firearms since their earliest days. To make the powder explode inside the gun, fire had to be introduced from without. External fire meant uncertain ignition and an inevitable delay between the pull of the trigger and the firing of the gun. Even with the best shotgun and fastest powder, hitting a bird on the wing was a challenge. Rain and wind added obvious difficulties for the shooter. Forsyth was determined to give the hunter a better chance. In doing so he radically transformed firearms technology.

Forsyth gave up Howard's idea of replacing gunpowder and went for a new type of primer. He first tried using the fulminate in the pan of his flintlock, but its flame didn't pass through the touchhole. Next he got rid of the flint and used the cock lever to deliver a blow directly to the fulminate—he was getting warmer. Finally he invented a device that could deliver a pinch of fulminate into a tube inserted into his gun's touchhole. The gun was known by the appellation "scent bottle" because the ingenious primer magazine on the side of the gun resembled a perfume vial. The cock hammered home a pin that set off this primer, shooting a jet of fire directly into the gunpowder. The delay between trigger pull and firing was dramatically reduced; the gun was no longer subject to the caprice of the weather. The system minimized misfires and reduced the escape of hot gas from the touchhole, adding force to the shot. The flintlock was doomed.

Over the twenty years that followed, inventors and tinkerers continually simplified and improved Forsyth's system. They shaped the fulminate into pills and packed it into copper tubes. They enclosed it between strips of paper, creating a detonator similar to the caps used in toy pistols. Around 1814, a little copper top hat charged with fulminate proved to offer the best solution. The cap fit onto a nipple projecting from the top of the gun's powder chamber. The hammer snapped onto it, igniting the fulminate, which set off the powder. The percussion cap was simple, waterproof, efficient.

The percussion system would have an impact far beyond the world of duck hunting. Once the approach proved reliable, armies began the rather simple procedure of converting flintlocks to the fulminate-fired weapons. Troops could fire at a faster rate, drill was simplified, and wet weather no longer turned a firearm into a club. The new system dramatically reduced the age-old problem of misfires. Percussion ignition made practical the revolver, the breech-loading rifle, and the repeater, further multiplying the firepower of a single shooter. The simplicity of the percussion system compared to the flintlock eased the transition from handmade to mass-produced firearms.

In 1817 a hunting enthusiast objected to the trend, protesting that the flintlock was an essential part of the sport's tradition. He answered those who touted the percussion cap's performance in the rain by noting, "Gentlemen do not go sporting in such weather." Then, with considerable prescience, he raised an objection that has echoed down the long history of gunpowder technology. "If, moreover, this new system were applied to the military, war would shortly become so frightful as to exceed all bounds of imagination." Such war, he wrote, would destroy "not only armies, but civilization itself."

———

WITH THE RISE of the Du Pont mills, America climbed to an equal footing with Europe in the production of gunpowder. As the nineteenth century progressed, the young nation began to take the lead in overall gunpowder technology away from the continent that had held it for four hundred years. One who contributed significantly to this trend was Samuel Colt.

Colt embodied qualities which came to be seen as typically American: abrasive, self-made, persistent, eminently practical in his thinking, as imaginative as he was mercenary, an opportunist, a liar, a genius. Beginning in the 1830s he tackled a problem that was, if anything, older than the one that had recently been solved by the Forsyth percussion system: how to let a shooter take more than one shot before reloading.

The vulnerability of a soldier during the complicated task of loading a firearm had been recognized by military theorists for centuries. Commanders had devised innumerable formations, drills, and tactics to compensate for the weakness. Leonardo da Vinci left behind drawings of "organ guns" consisting of multiple barrels, unwieldy contraptions that needed a carriage or framework for support. Gunsmiths devised many other varieties of multi-barreled handguns during the flintlock era, but few proved practical. Forsyth's invention inspired new attempts to crack the problem.

Sam Colt's father was a failed New England entrepreneur and the family had bounced in and out of poverty. Born in 1814, Sam was apprenticed to a farmer at age 11 and spent time working in a Ware, Massachusetts, textile factory. He nurtured an abiding love of gunpowder. As a Fourth of July stunt, the 15-year-old advertised that he would blow up a raft in Ware Pond using underwater explosives. He missed the raft, but sent an impressive geyser skyward. Trundled off to boarding school, he fascinated classmates with his pyrotechnics. When the Fourth rolled around again he accidentally set a fire that ended his formal education.

Now it was off to sea for the enterprising lad. During a trip to Calcutta he mulled the notion of a multi-shot handgun. Supposedly inspired by the way the spokes of the helmsman's wheel lined up with a clutch, he whittled a wooden model of a revolver. The gun would have a single barrel; multiple breech chambers holding powder and ball would turn to line up with it. Back home his father hired an unenthusiastic gunsmith to make a model that shattered with the first shot.

"Once again I am on the move to seek my fortune," Sam said, an apt summary of his life.

From an amateur chemist at the Ware factory he had learned how to make nitrous oxide, known as laughing gas. He started with a handcart, giving demonstrations of the harmless intoxicant on the street. By the time he was 18 he had transformed himself into "Dr. Coult, of New York, London and Calcutta." He moved his act into opera houses and auditoriums, breathing the gas himself and letting audience members partake. He touted the show as a scientific demonstration, but the few minutes of giggling and babbling that the substance induced provided ample hilarity for the audience.

Colt had by no means lost sight of his goal. His laughing gas tour was intended only to bankroll his revolver. It gave him capital—he was earning $10 a day at one point—and something just as valuable. His short career as a performer left a lasting appreciation of the value of showmanship. Colt established himself as one of the giants in the long history of gunpowder weapons not only because he was a technical genius, but also because he made himself a master of ballyhoo.

Having nailed down patents in England and America, Colt managed to raise $230,000 in capital from well-heeled relatives. He set up a factory in Paterson, New Jersey. Each chamber of the revolvers he produced had to be loaded individually from the front. The shooter then fixed a primer cap on a nipple at the rear end. Cocking brought a chamber into position behind the barrel. Pulling the trigger let the hammer snap against the cap. The mercury fulminate inside, mixed with other flammable chemicals, sent a jet of flame into the chamber, blasting the bullet down the barrel.

The newness of the revolver concept and the opposition of a handful of government officials, though, proved stumbling blocks to acceptance of Colt's ideas. In the American military, political cronyism was rampant, resistance to innovation a reflex. When Colt landed a trial at West Point, the Army judged that the revolver's "complicated character, its liability to accident and other reasons," made it unsuited for service. British authorities refused even to test the weapon.

Colt did manage to sell a few revolvers that found their way to the newly independent Republic of Texas. The gun became a popular weapon in the ragtag militia known as the Texas Rangers. Colt kept tinkering with and improving his invention. He reduced the number of moving parts from thirty-six to twenty-eight, and finally to seven. But failure to secure substantial government contracts forced him into bankruptcy. In 1842, at age 28, Colt was broke.

With his long-ago demonstration at Ware Pond in mind, he turned his attention to blowing up ships with mines. He used electric batteries to set off floating barrels of gunpowder. Always the showman, he sank a 500-ton schooner in the Potomac River while government bigwigs and a large crowd looked on. Colt claimed he could "destroy a hostile fleet without exposing the life of a single American." John Quincy Adams, the former president, objected vehemently. "Killing men by infernal contraptions on the bottom of the sea," he stated, was both uncivilized and "unchristian." This debate, between the technologist and the moralist, had surfaced often during the history of gunpowder. The government, for reasons of its own, declined to pursue the idea.

By then, the Colt revolver was beginning to show itself a decided improvement over rival arms. Colt collected stacks of letters from officers who had actually used the pistols in combat and praised them profusely. The guns were especially valued for fighting from horseback—a small group of soldiers, each armed with a brace of Colts, could delivering a withering storm of bullets.

The outbreak of the Mexican war in 1846 allowed Colt to land his first substantial contract for pistols. Lacking a factory, he had to subcontract the work. As the firearms proved themselves during the victorious American campaign, sales began to mount.

Colt's fortunes changed almost overnight. He established a factory in Hartford in 1847. Two years later he managed to renew his patents and make his first profits. He began producing a handy .31 caliber pocket revolver—he would sell 325,000 of these pistols during his lifetime. Grizzled gold miners and city sharps alike all wanted a Colt revolver.

Colt's influence on gunpowder technology extended beyond his solution to the multi-shot problem. Until the 1850s, guns had been made, for the most part, by skilled artisans, who shaved and bored stocks by hand, forged barrels by hammering them on anvils, assembled each lock as a custom mechanism. Firearms were expensive, production slow, repair in the field virtually impossible.

Colt adopted a system, pioneered by Eli Whitney a half-century earlier, that used machine tools to produce interchangeable parts. The concept was similar to that of a modern key-cutting machine. A prototype guided the cutting blade to reproduce the original form exactly. With the turning, boring, and milling machines available at midcentury, Colt could transform the ancient craft of the gunsmith into a form of mass production.

What came to be known as the "American System" of factory production was not Colt's invention, but he contributed mightily to the system's widespread adoption. In 1851 he attended the International Exposition in London's Crystal Palace. He placed the parts from disassembled revolvers in bins, mixed them up, and built a functioning gun on the spot from randomly selected parts. The guilds of gunsmiths vehemently opposed his system, and for good reason. The vulgar and presuming American had bested the English "in an art which they had practiced and studied for centuries." The American System allowed semiskilled workers to shape parts to fine tolerances. It took the skill out of the hands of the craftsman and put it into those of the machinery designers and, ultimately, the financiers.

Colt made a grand tour of the world's capitals. Granted an audience with the Ottoman sultan, he presented him with a pair of ornately engraved pistols and let slip the fact that the Russians were rushing to buy revolvers. He failed to mention that his Russian orders were the result of similar hints about the Turks that he had dropped at the Czar's court.

Though the most famous Colt revolver, the "Peacemaker," was designed after his death, Sam always believed that his invention was an agent of justice and tranquillity. A gentleman "armed with my invention can keep a dozen ruffians at bay," Colt proclaimed. If he ignored the fact that ruffians and malefactors could increase their own fire-power with Colt weapons, who could blame him? There was money to be made.

Colt became fantastically wealthy. He made his factory in Hart-ford a model of nineteenth-century labor relations, with a company-sponsored social hall, art exhibits, and band. Unfortunately, he did not enjoy his riches for long: He died in 1862 at the age of 48.

———

COLT'S SUCCESS pushed his country to the forefront in the rapid modernization of the arms industry. The career of another firearms entrepreneur, Richard Jordan Gatling, emphasized the particularly lively enthusiasm with which Americans pursued gunpowder technol-ogy. Four years younger than Colt, Gatling had invented a number of successful farm machines, including a steam plow.

With the arrival of the Civil War, inventors went wild devising new armaments. A writer for the *Philadelphia Enquirer* in 1861 was sure that Yankee genius would produce some "patent Secession-Excavator, some Traitor-Annihilator, some Rebel-Thrasher." The Chief of Army Ordnance was so deluged with crackpot proposals that he stopped considering further submissions.

"It occurred to me," Gatling recalled later, "if I could invent a ma-chine—a gun—which could by its rapidity of fire, enable one man to do as much battle duty as a hundred, that it would, to a great extent, supersede the necessity of large armies."

This idea of a weapon as a labor-saving device was typically Ameri-can and, despite its apparent naivety, very modern. It was the idea of a man who had never been to war, of the "gentlest and kindliest of men"—so said Gatling's obituary in *Scientific American*. It was the idea of a man gripped by the pervasive nineteenth-century concept of the ul-timate weapon. "By making war more terrible," the magazine noted, "it seemed to him nations would be less willing to resort to arms."

The Gatling gun, which used multiple revolving barrels to fire 200 gunpowder cartridges a minute, was one of the most successful of the early machine guns. It carried Colt's idea to another level.

Though Gatling demonstrated his weapon to the Union Army in December of 1862, six months before Gettysburg, the Ordnance Department did not accept it until after the Civil War had ended. It was later deployed on both land and sea. In large calibers it offered an effective defensive weapon for ships. Its intimidating array of barrels and reputation for lethality also made it a useful riot gun. The British used Gatling guns to suppress an uprising in Port Said, Egypt, in 1883. "This gun is a wonderful 'peacemaker' in such cases," the *Daily Intelligencer* noted.

Other machine guns would follow, including the French *mitrailleuse* of the 1860s and the Maxim belt-fed automatic gun of the 1880s, ratcheting firepower to unprecedented levels. Gatling was correct that the weapons made war more terrible. He erred in thinking that the horror would diminish nations' willingness to fight.

———

IN 1823 A British officer named Colonel Norton, stationed in South India, observed natives shooting blowguns. He described how they attached to each dart a round disk of lotus pith. When they blew into the mouthpiece, the pith expanded, sealing the tube and accelerating the projectile. French Captain Claude-Etienne Minié of the Chausseurs d'Orleans took up this idea to solve an ancient dilemma. Infantrymen had been forced to choose between the quick-loading, inaccurate musket and the slow, precise rifle. Minié designed a conical bullet narrow enough to drop easily down the bore, but with a hollow base that widened with the blast of the powder. The bullet caught the grooves of the gun's rifling, which imparted spin to the projectile.

Besides taking advantage of rifling, Minié's elongated bullet presented a more aerodynamic shape to the air and therefore encountered less resistance. It was able to translate the explosive force of gunpowder into hitting power at the target with less energy loss. The spherical projectile, in service since the days of the earliest gun stones, became obsolete overnight.

No military commander could fail to see the advantage of equipping troops with fast-loading rifles that could shoot three or four times as far as smoothbore muskets with greater accuracy and power. Converting to the Minié ball was relatively simple: Muskets only needed to be bored with rifling grooves—old weapons could easily be rebored to work with the new ammunition. European nations rushed to embrace the system. In America, Secretary of War Jefferson Davis recommended adoption of what would be known as the "minnie" ball in 1854. Smoothbore musket production soon ceased.

The normally sluggish British military establishment bought the rights to Minié's invention in 1851 and improved on it. Four years later they established a plant in Enfield using the American system of manufacturing and began mass-producing some of the best firearms in the world. The Enfield rifle used a traditional paper cartridge containing bullet and powder. The paper was heavily greased to deter moisture and to provide a lubricant that eased loading. All the shooter had to do was to bite off the end of the paper to expose the gunpowder inside and ram the whole thing down the barrel.

This simple procedure became the spark for one of Britain's most fiercely fought colonial wars, a struggle that was termed, with Victorian hyperbole, an "epic of the Race." For 150 years the British East India Company had ruled the Indian subcontinent, its power resting on an army of native soldiers, or "sepoys," commanded by British officers. With the French driven out and native resistance crushed, 1857 saw British control of the vast territory unchallenged. That year a rumor spread among the sepoys that a combination of fat from both pigs and cows was used to grease the new rifle cartridge—the former were abhorrent to Muslims, the latter sacred to Hindus. Biting the cartridges became an act of sacrilege.

The lubricant used on the cartridge did in fact contain beef tallow and in some cases hog grease. No offense was intended, however. British officers felt that the issuing of these modern weapons was a sign of trust in their dark-skinned subordinates. Orders were quickly passed down to replace the offending grease with an oil-beeswax mixture. It was too late.

When a troop of cavalrymen in Meerut, north of Delhi, refused to handle the new cartridges, they were publicly stripped of their uniforms, shackled, and sent off to hard labor—some of them had served the British in India for thirty years. The next day a band of sepoys, maddened by the humiliation, broke into the jail, freed their comrades, and led a mob on a rampage. They murdered many Europeans, then fled to Delhi, carrying with them the seeds of mutiny. A year of horrendous violence ensued. The enraged British invented a punishment for mutineers known as the "devil's wind." They lashed the man over the muzzle of a cannon and fired a ball through his body, blowing him to bits and demolishing his hope of an afterlife. The war ended with the establishment of direct rule from London—the Raj—and dissolution of the East India Company.

In the coming decades, the paper cartridges that had touched off the rebellion became obsolete, completing another advance in gunpowder technology. Copper or brass cartridges containing bullet, gunpowder, and primer were not just easy to handle, they opened new avenues of progress in firearms. Their principal contribution was that when the gun was fired, the soft metal expanded, sealing the breech. Shooters could load their weapons from the rear without worrying about the leakage of hot gases.

Various inventors tried approaches to making metal cartridges work. They settled on a brass cartridge with the primer built into its center, a bullet attached at the other end. This invention proved essential to the development of efficient repeating rifles. New England gunsmiths Horace Smith and Daniel Wesson first put the metal cartridges into revolvers, obviating Colt's original system of loading each chamber from the front. Christopher Spencer, a Connecticut inventor, designed a rifle in 1860 that used a lever to open the breech and load one of the seven cartridges lined up in a hollow in the stock. His and other repeaters found limited use in the Civil War.

With the arrival of the metal cartridge, almost all the pieces were in place to create the modern weapon. The history of firearms, which stretched back to the hand cannon of the fourteenth century, would soon be complete. The shooter of the twenty-first century continues to use a gun almost all of whose technology was invented before 1870. One final improvement remained: the replacement of the gunpowder itself with a more powerful synthetic propellant.

———

WHILE SMALL ARMS were reaching a pinnacle of development, an accident in 1844 offered dramatic proof that in large cannon gunpowder's wild nature had yet to be tamed. The incident pointed the way toward one of the last important changes in the way gunpowder was made and used.

Captain Robert F. Stockton was an ambitious naval officer. Born in 1795, he dropped out of college to go to sea. He turned down President John Tyler's offer to appoint him Secretary of the Navy, preferring active duty. He pursued a vision of a modernized Navy made up of armored steam ships mounting large guns. The embodiment of his dream was a ship christened the *Princeton* after his home town.

This vessel straddled two eras. It was fully rigged with sails, as battleships had been since the age of Henry VIII. It was also one of the first warships with propulsion aided by a screw propeller and a steam engine set below the waterline. Suddenly the maneuverability and self-propulsion of the ancient oared galley had returned to naval warfare.

Contributing to the design of both the ship's propulsion system and its armament was John Ericsson, the Swedish engineer who had invented the screw propeller and who would go on to build the *Monitor*, the turreted armored warship that the U.S. Navy would introduce during the Civil War. Forty-two-pounder guns were considered formidable at the time, and Ericsson supplied the *Princeton* with a dozen of them. But the ship's two main guns were even bigger, their gaping 12-inch bores designed to fire solid cannonballs weighing 212 pounds. It was a return to the age of the bombards, and like bombards, these guns were large enough to have their own names.

Though the *Princeton*'s guns were far bigger than previous naval armaments, Ericsson and Stockton had faith in a new concept of gun making. Ordinary cast iron had proven too brittle for really large guns. To forge the ship's weapons Ericsson went back to wrought iron, the material used in fifteenth-century bombards. He bored the barrel from a long chunk of this material. Stockton tested the first gun, the Oregon, by firing it with a full load of powder. When he noticed a small crack, he ordered that two thick wrought iron sleeves be shrunk around it. To strengthen the second gun, Stockton welded on an additional one-foot thickness of iron at the breech. A new age of naval weaponry was at hand.

The *Princeton* was fitted out in New York. Regarding the second gun, the watch officer recorded, "they christened it the Peace Maker with six cheers." The name, with its suggestions of both serenity and menace, was a favorite of weapon makers. Of the *Princeton*'s armament, Stockton boasted, "It is worth all the guns on board of any frigate."

In Washington, President Tyler inspected the ship and immediately recommended that Congress authorize several more. Stockton sent out engraved invitations to a gala cruise and banquet. On February 28, 1844, more than 500 ladies and gentlemen crowded aboard, including congressmen, cabinet members, and diplomats. Tyler was in attendance, along with the 24-year-old Julia Gardiner, who would soon become his second wife.

The shooting of the Peacemaker was an essential part of Stockton's spectacle. That day he fired the big gun twice, astonishing his guests with its gargantuan roar. Then all retired below to feast, drink toasts, and listen to speeches. Later in the afternoon, someone asked Stockton to demonstrate the gun once more. While the party continued, the captain and a sizable group went up on deck.

A witness reported that the sound of this third firing was curiously muffled. The shot was followed by an ominous stillness during which the laughter and hubbub of the festivities below could be heard. Then came screams and the urgent bark of orders. Stockton was led below by two sailors, "his full black wig . . . blown away, and his head bound up in cloths saturated with blood." The gun had exploded.

Among those killed by flying fragments of iron were the Secretary of State, the newly appointed Secretary of the Navy, several other government officials, two sailors, and Julia Gardiner's father.

News of the tragedy riveted the nation. What did this mean about the future of the Navy? About America's impending showdown with Britain in the disputed Northwest? Tyler ordered the victims laid in state in the White House. He tried to cover his own reputation by blandly describing the mishap as "one of those tragedies which . . . are invariably incident to the temporal affairs of mankind." In a sense his phlegmatic judgment was accurate. The government luminaries had learned what gunners had known for centuries: gunpowder was capable of turning on its user without warning.

A court of inquiry cleared Stockton of all blame—during the Mexican War he gained renown as the conqueror of Los Angeles and later served as a United States senator. Knowledgeable observers knew, though, that the casual construction and freewheeling demonstration of the big gun was a flagrant bit of hubris. Modern guns were too massive, modern powder too strong, to leave ordnance design to empirical, common-sense analysis. A more rigorous and systematic approach was called for.

———

THE SPECTACULAR accident aboard the *Princeton* inspired the career choice of the man who would make the biggest contribution to the development of gunpowder in the nineteenth century. Thomas Jackson Rodman was not a pure scientist but a soldier-technologist in the tradition of the ancient gunners. Born in Indiana in 1815, he attended the Military Academy at West Point, where he showed a particular aptitude for mathematics and mechanics. He graduated in 1841, served as an ordnance officer during the Mexican War, and then pursued problems related to artillery. Large guns would clearly have a role to play in any future war; and large guns just as clearly presented technical problems of the most challenging complexity.

Rodman's first step in designing an artillery piece that was both large and safe was to create gauges that could accurately measure the tremendous pressures generated along the barrel. Building on what was known about metallurgy, he devised a new method of gun casting that addressed weaknesses that had plagued gunners for centuries. The original cast bronze guns had been formed around a core, which, when removed, left the hollow tube. In the middle of the 1700s, Swiss gun casters invented a way to cast cannon as a solid block, then drill out the bore, producing a stronger piece and a truer alignment.

Rodman turned back the clock. He proposed casting the gun around a central core cooled by circulating water. The iron around the bore would harden first. As the outer layers solidified, they would, according to the theories of metallurgy, shrink around the inner layers, putting them under permanent tension. The gun would be stronger, and the stress of the explosion would be more evenly absorbed by all the metal.

The government first looked on Rodman's idea as too radical, but tests began to indicate that he might be right. Rodman supervised the casting of a prototype in 1859. He fired more than 500 proofing rounds, establishing that he had fashioned the strongest large cannon ever built. His systematic, informed approach to the problem reduced the likelihood of another disaster like that on the *Princeton*.

With war clouds gathering, the federal government ordered Rodman to cast a gun with a 15-inch bore. The "Lincoln Gun" had a bottle-shaped barrel 16 feet long that weighed 25 tons. The cannon used 130 pounds of powder to fire a 440-pound iron ball more than three miles.

Even as he created his massive new gun, Rodman was conducting a systematic appraisal of the powder it would burn. The "best" gunpowder had long been viewed as powder that was most explosive. In fact, there was no single best powder. The force needed to accelerate the enormous projectiles of heavy guns was not a sudden burst of pressure, but a sustained push. Slower, not faster powder was called for.

With guns of enormous proportions like Rodman's, the problem grew critical. Big cannon burst more frequently and wore out more quickly than small guns, but gunpowder had not been seen as the source of the problem. It had long been known that slightly larger grained powder worked better in larger pieces, but no one had considered the issue more carefully—all cannons of the time used the same powder.

Rodman assumed that large-grained powder would burn more slowly, thereby reducing the initial pressure. "As for our common cannon powder," he said, "it is too fine grained and too explosive even for field guns, and certainly never should be used for guns of large caliber."

He took his idea to Lammot du Pont, the chemist of the powder making firm. They tested powder with grain sizes up to half an inch. They found that these largest grains dramatically reduced peak pressure with only small loss in the velocity of the projectile. In the 15-inch Rodman gun, the powder produced a peak pressure only 20 percent as high as that of ordinary cannon powder. This "Mammoth" powder was chosen as the standard. It was made from high-density presscake broken into grains averaging six tenths of an inch in diameter, then rounded off in the glazing barrel in the usual way. The propellant was no longer a powder but resembled small pebbles.

With his new powder and new method of casting iron, Rodman felt confident that he could produce a gun of almost any size. With war raging, the government ordered a super-gun to protect New York

Harbor. It took Rodman three years to design the weapon and build the mold. Six furnaces smelted the iron. Oversized tackle and lathes handled and finished the piece. The barrel weighed 58 tons, its cavernous bore stretched 20 inches across. "Juveniles . . . were amusing themselves today crawling into the bore on their hands and knees," the *Pittsburgh Gazette* reported. Mehmed the Conqueror would have been envious.

A special rail car transported the gun to Fort Hamilton in Brooklyn, where workmen installed it in an 18-ton carriage. On a day in October of 1864, with a crowd looking on, Rodman loaded the piece with a hundred pounds of Mammoth powder and a 1,080-pound cannonball. When he fired the gun, the massive projectile exploded out the muzzle accompanied by a cloud of smoke as big as a house and a monstrous roar. The ball landed three and a half miles out at sea.

Rodman had used a mixture of disciplined science and shrewd engineering to solve long-standing problems. Like the Du Ponts, Colt, Gatling, and other ingenious pioneers, he had brought the ancient explosive to a new peak of efficiency just in time for the single largest conflagration of gunpowder in history, the American Civil War.

12

APPALLING GRANDEUR

AS VOLUNTEERS departed to fight in the American Civil War, many had their photographs taken and made into *cartes de visite* for their friends and sweethearts. The pictures show their uniforms neat and clean, their faces confident, proud, and eager. "So impatient did I become for starting that I felt like a thousand pins were pricking me," an Arkansas man wrote. Washington socialites and congressmen were equally giddy about the prospect of war. They carried picnic hampers and champagne as they rode out in their carriages to watch the conflict's first big battle near Manassas Junction, Virginia.

Neither participants nor spectators could imagine what lay ahead, in the battle or the war. By the 1860s, more effective weaponry was available than had been used in earlier wars. The rifled musket, taking advantage of the Minié ball, put fire of unprecedented accuracy in the hands of each soldier. Gunpowder itself had nearly reached its most powerful and advanced state of development.

Generals were slow to adjust their tactics to the newly lethal battlefield. Inspired by the gallantry of the previous century and by the shock tactics with which Napoleon had temporarily dominated Europe, officers urged their men to counter the roar of gunpowder with renewed courage. But courage, as many brave soldiers soon learned, could no longer overcome firepower.

Though the tensions that gave rise to the conflict had been mounting for decades, the combatants had neglected to put their gunpowder supplies on a firm footing. "Neither party," Lincoln would note four year later, "expected for the war, the magnitude, or the duration, which it has already attained." Most estimated that the conflict would be decided in six months.

The Union states were better supplied with gunpowder than the Confederacy. Dozens of mills were scattered across the North. Some of them, like the Oriental Powder Company in Maine and the Schaghticoke Powder Company outside of Albany, New York, were substantial industrial facilities. The Du Pont works itself turned out nearly half the nation's gunpowder. The huge powder mill in Wilmington made the question of Delaware's loyalty to the Union a critical issue. Though they supported slavery in their state, the citizens of Delaware rejected calls to secede. Henry du Pont declared his allegiance to the Union and refused to sell gunpowder to the states that chose to break away. The proximity of his mills to Confederate territory would remain a worry throughout the war.

The main obstacle facing Du Pont as he prepared to meet the enormously expanded demand for gunpowder was a shortage of saltpeter. The United States imported most of this key ingredient from India through British dealers. Sympathy with the Confederacy was strong in Britain. That country's government was debating whether to recognize the Confederate States of America, or even to intervene on the Southern side.

Henry du Pont took immediate action. He sent to England his 30-year-old nephew Lammot. Holding a degree in chemistry from the University of Pennsylvania, Lammot was one of the most brilliant of all the Du Ponts. In a single day in November 1861 he bought every ounce of saltpeter that was available in England and contracted for further shipments on their way from India. In all, he acquired 3.4 million pounds of the vital commodity. Ostensibly, he was buying this supply for the Du Pont partnership. In fact, his trip had been authorized by the Secretary of War, and federal funds had been slipped into Du Pont accounts in London.

Du Pont mills began working around the clock, operating perilously by lantern and candle light. The fatigue and carelessness induced by the rush contributed to the danger. The fear of sabotage was constant. During the war, eleven separate explosions tore through the works. Forty-three men died. The mills were rebuilt, the work went on.

The demands of the war spurred ingenuity. By 1863 saltpeter was again running low. The price of Indian saltpeter had gone through the roof. Company chemists devised a method to convert Chilean saltpeter, which was predominantly sodium nitrate, to the potassium nitrate needed for gunpowder. At first they used natural potash, then improved the process by mixing in a solution of potassium chloride. Eventually, this chemical process would replace the laborious chore of extracting and refining natural saltpeter, breaking the centuries-old British monopoly on Indian saltpeter.

————

WHEN THE GUNS OPENED on Fort Sumter, the powder supply in the South was tenuous. With only four small mills, each producing less than 500 pounds of gunpowder a day, and with only 491,000 pounds of the explosive on hand, the Confederacy faced a predicament like the one that had afflicted the thirteen colonies at the opening of the American

Revolution. Unless the situation could be rectified, the secessionist cause would wither for lack of gunpowder.

Faced with this bleak prospect, the Confederacy turned to George Washington Rains. Born in rural North Carolina in 1817, Rains had excelled at West Point and had taught chemistry there, resigning his commission in 1856 to become president of an iron works in nearby Newburgh, New York. Though he had lived half his life in the North, Rains chose allegiance to the slave states when war erupted. Confederate president Jefferson Davis put him in charge of what would soon be called the Gunpowder and Niter Bureau.

Rains knew that local producers could supply plenty of charcoal—he determined that cottonwood, more plentiful in the South than willow, worked just as well as a powder ingredient. Several hundred tons of sulfur, imported for sugar refining, were on hand at New Orleans. Adequate additional supplies could be acquired from sources in Texas.

Saltpeter was the critical item. Rains dispatched agents to Europe, and over the course of the war smugglers brought 2.7 million pounds of the salt through the blockade by which the Federals intended to strangle their enemies. Deposits derived from bat guano were dug from caves in Tennessee, Kentucky, and Alabama. To supplement these sources and to assure Confederate self-sufficiency in niter, Rains turned to the age-old practice of establishing saltpeter "plantations" or "nitriaries." Workmen dug long pits and filled them with stable manure, rotting vegetation, and animal carcasses. Stray dogs were rounded up and tossed in, prompting the *Montgomery Weekly Mail* to quip that "soldiers using this powder are said to make a peculiar *dogged* resistance."

Jonathan Haralson, a zealous official of the niter district around Selma, Alabama, insisted that housewives save the contents of their chamber pots to add to a collection barrel. Confederate soldiers composed a bit of doggerel to commemorate the proposal:

We thought the girls had worked enough in making shirts and kissing,

But you have put the pretty dears to patriotic pissing.

Not to be outdone, the Yankee troops replied:

No wonder that your boys are brave! Who couldn't be a fighter,
If every time he shot his gun he used his sweetheart's nitre?

The Confederacy fell before much of this homegrown saltpeter, which took at least 18 months to ripen, found its way to rebel guns.

The task facing George Rains was anything but a joke. He had to create an efficient gunpowder industry from scratch. "Without plans, without machine shops, without powder makers, without mechanics," he said, "I was required to erect somewhere a giant works. I was thrown upon my resource to supply these deficiencies."

His guide was a pamphlet written by a British artillery officer describing the up-to-date government powder plant at Waltham Abbey in England. With advice from a man who had once worked at that mill, Rains oversaw the construction of the most modern gunpowder factory in the world in Augusta, Georgia. He laid out the plant for efficiency, with the raw materials entering at one end, passing through each step of the process, and emerging as finished powder a mile and a half down the line.

Rains chose Augusta because it was a secure inland location with good rail and water transportation. He obtained a 130-horsepower steam engine and connected it to a dozen incorporating mills, each with a pair of 5-ton wheels. For safety, Rains added a drencher device used in England. If a mill caught fire a gush of water automatically poured down on all the grinding wheels to prevent a chain reaction.

Once the plant was running at full capacity, it ground three and a half tons of gunpowder a day, 2.75 million pounds over the course of the war. The Augusta works were among the few permanent structures erected by the hard-pressed Confederates. Rains had had the giant refinery building designed as a replica of the British House of

Parliament. That a nation fighting for its existence and strapped for resources would create an industrial plant in the shape of a medieval castle suggests something of the romanticism at the core of the Confederate experience.

———

CONFEDERATE President Jefferson Davis hoped for quick victories that would sap the will of the Union to fight. He also needed a defensive strategy to counter the North's "Anaconda Plan," a naval blockade aimed at crushing the rebellion. To forestall this scheme, the Confederates put gunpowder to work in a class of weapons whose use inflames controversy even today.

During the Civil War, the word "torpedo" meant a device containing a charge of gunpowder and intended to sink or disable a ship, or a similar type of buried explosive that we would call a land mine. The idea was not new. Mines had been used in China as early as the thirteenth century. The steamboat inventor Robert Fulton had experimented with torpedoes, and they had been used in the recent Crimean War.

The question of whether such a device was an ethical means of waging war was not a settled question. The torpedo and related weapons were sneaky and indiscriminate. Gunpowder had already expanded the distance at which a man could kill. Some thought that torpedoes made war unacceptably mechanical, anonymous, and inhuman. Like guns before them, they were disparaged as the tools of cowards, offenses against decency and civilized warfare.

The man most responsible for this ingenious and remarkably effective Confederate use of gunpowder was General Gabriel Rains, older brother of the South's explosives wizard. That both men were involved with gunpowder was something of a coincidence: Fourteen year apart in age, they had scant personal relationship and never worked directly together.

In the spring of 1862, with the Confederates retreating from York-town, Virginia, the elder Rains commanded an ineffectual rearguard. To gain time he ordered his troops to bury 8- and 10-inch artillery shells with detonators attached "simply as a desperate effort to dis-tance our men from the pursuing Union cavalry." The shells exploded and whole companies of Yankees bolted in panic. Union general George McClellan roared that "the rebels have been guilty of the most murderous and barbarous conduct." General James Longstreet, Rains' superior officer, forbade the further use of gunpowder in this manner. He condemned the mines as not a "proper or effective method of war." Longstreet's view did not win out. Rains was put in charge of a broad program that would make use of torpedoes, mines, and similar gunpowder devices.

The urgent military situation spurred Southerners to the re-sourcefulness of the desperate. "Many an ingenious mind turned its attention to . . . inventing some *machine infernale*," a contemporary observer noted. The Confederate war department, like its northern counterpart, was plagued by proposals, particularly after the gov-ernment offered a reward for any ship sunk or Union facility ruined. Inventors envisioned torpedo boats powered by rockets, diving ap-paratus for attaching explosives to ships, balloons for dropping bombs on enemy targets. A man named R. O. Davidson proposed a "Bird of Art." This was "a machine for aerial locomotion by man" carrying a 50-pound load of exploding shells. A thousand of the birds, he was sure, would put an immediate end to the war. He called on every southern patriot to send him a dollar so that he could start building them.

Gabriel Rains was more practical. He focused on land mines and on mechanically operated floating torpedoes, which were touched off by contact with a ship's hull. The fulminate that Reverend Forsyth had used to ignite his fowling piece offered an ideal substance to in-corporate into a torpedo fuse. A hard knock would set off the primer, which would transmit flame to the main powder charge.

Rains countered arguments that this method of warfare was ignoble by answering that it was justified in "defense against an army of Abolitionists, invading our country." His confidence in the technique was as boundless as his hatred of Yankees. "No soldier will march over mined land," he asserted. A corps of sappers armed with mines "could stop an army."

The justification of torpedoes resulted in some fine distinctions. "It is admissible to plant shells in a parapet to repel assault, or in a road to check pursuit," Confederate Secretary of War George Randolph concluded. "It is not admissible to plant shells merely to destroy life and without other design than that of depriving the enemy of a few men."

Among those with mixed feelings was Confederate Lieutenant Isaac M. Brown. He had helped set up the defenses of a makeshift shipyard on the Yazoo River in Mississippi. In December of 1862, his mines blew up the U.S. ironclad *Cairo,* the first ship sunk in combat by an electrically detonated torpedo. Brown said he felt "much as a schoolboy . . . whose practical joke has taken a more serious shape than he expected."

The most famous incident of torpedo warfare took place at Mobile Bay on August 4, 1864. Though a native of Tennessee, 63-year-old Admiral David Farragut had remained loyal to the Union. His actions in the Gulf of Mexico and on the Mississippi had contributed to the capture of New Orleans and Vicksburg. The tightening naval noose had reduced the ports available to southern blockade runners to a handful. Farragut was determined to erase Mobile from the list.

The rebels had floated numerous torpedoes in the harbor mouth, leaving a channel protected by the guns of Fort Morgan. Farragut directed a flotilla of four iron-clad monitors and fourteen wooden warships up the channel, led by the armored *Tecumseh.* He loathed the hidden weapons as unworthy of a "chivalrous nation." As he watched from the rigging of his flagship, the *Tecumseh* steered out of the channel and exploded a torpedo. Its propeller rose from the water still turning and in two minutes the ship had disappeared, taking 120 men from a crew of 141 to the bottom. More torpedoes were spotted in front of the squadron. Indeed, the ships' crews could hear detonators snapping against their hulls. The admiral had already decided to press ahead, forcing as many ships through as he could. "Damn the torpedoes!" he shouted. "Four bells! Captain Drayton, go ahead! Jouett, full speed!"

The Union forces continued on and captured the fort that dominated the mouth of the bay. The city of Mobile held out, but its usefulness as a port had ended.

———

AS THE WAR grew more desperate, the Confederates turned increasingly to saboteurs. "Boat-burners" became a menace on the Mississippi. They spiked firewood and lumps of iron shaped like coal with gunpowder charges and tossed them into fuel bins, hoping to disable Union vessels.

Union General Ulysses S. Grant witnessed one of the most dramatic sabotage operations of the war. On August 9, 1864, with his army bogged down in the long siege of Petersburg, Virginia, he was sitting outside his tent on a bluff overlooking the James River at the main Federal supply depot of City Point. The weather was oppressive. Down on the wharfs sweating black stevedores were unloading supplies.

Confederate saboteur John Maxwell casually strolled into the busy depot along with his friend R. K. Dillard. He carried with him a box containing a bomb he had invented: 15 pounds of gunpowder, a detonator, and a clockwork device to set it off. He called it a "horological torpedo." Around 10 A.M. the two men passed the box to a worker on a barge, saying the captain had ordered it aboard. They strolled away.

Around midday the bomb exploded, setting off the artillery shells and small arms ammunition on board the barge, spreading quickly to other loads of gunpowder on the wharf. One newspaper reporter described it as "a stunning concussion . . . accompanied by a sound

which some compare to the discharge of a cannon close to each ear."
Another witness simply said it was "staggering." The barge disappeared. A mushroom cloud that could be seen for miles spread into
the sky. The shock wave knocked down buildings and wrenched ships
from their anchors. Cannon and horses flew through the air. Bullets
and shells shot wildly in all directions. A woman on a passenger ferry
reported that a man's head fell on the deck at her feet. "I picked it up
by the hair," she said, "and placed it in the deck bucket of water."
Grant maintained his sangfroid but later wrote to his wife, "It was
terrible—awful."

The blast killed at least 58 persons and wounded 126. More than
$4 million worth of supplies and equipment were destroyed. A court
of inquiry judged the incident an accident. It wasn't until after the war
that Maxwell's account surfaced.

———

THIS SIEGE AT Petersburg, twenty miles to the south of the Confederate capital Richmond, ate up the last year of the war. Its static
techniques brought to mind the warfare of earlier centuries, its
trenches and bombproofs were an omen of the disastrous fighting that
would engulf Europe fifty years later. Union artillery shelled Petersburg in an effort to dislodge the rebels. The city residents who had
stayed behind retired to their basements for shelter. As the shells flew,
a Mississippi soldier reported, "the little boys watch for them to fall
& if they don't explode they take out the powder & sell it."

The Federals' frustration grew. Lieutenant Colonel Henry Pleasants, a mining engineer before the war, suggested an idea as old as
siege warfare itself: digging under the Confederate lines. His superior,
General Ambrose Burnside, assigned him a crew of Pennsylvania coal
miners to do the digging. The workers bored into the hillside at a
point where the rebel and Union lines ran close together. They dug a
narrow tunnel, four and a half feet high and 500 feet long.

On July 27, miners packed the tunnel, twenty feet below the Confederate position, with 320 powder kegs, totaling 4 tons of explosive. The attack was to come the next day. After the explosion, Federal soldiers would storm through Confederate lines and seize high ground to the rear, opening the way to Richmond, and, it was hoped, ending the war. At 3:30 A.M., exactly on schedule, Lt. Col. Pleasants lit the fuse. "Hastening to the surface he stood with watch in hand," a subordinate recorded. "The time for the explosion passed." Nothing happened.

Pleasants went wild. He had already complained of the inferior fuse that he'd been supplied—short lengths that he had had to splice. Two men rushed into the tunnel and found where dampness had extinguished the fuse. They cut it, lit it again, and "regained the outside as rapidly as possible."

"First there came a deep shock and tremor of the earth and a jar like an earthquake," wrote a soldier of the 20th Michigan regiment. Then a "monstrous tongue of flame shot fully two hundred feet into the air followed by a fast column of white smoke." Earth sprayed skyward, "mingled with men and guns, timbers and planks, and every kind of debris."

"The earth seemed to tremble," an Alabama officer noted, "and the next instant there was a report that seemed to deafen all nature." Of the 300 South Carolina boys who manned the trenches over the mine, almost all were killed or mangled. A stunned Confederate soldier reported to his officer: "Hell has busted."

Northern troops began to charge through the huge crater that the blast had left. The scene, one said, "beggared description." The pit was littered with half-buried cannon, wounded men, body parts. The gunpowder smoke still seeping from crevices lent an infernal element. Though Pleasants had wanted to use even more powder, the explosion had proven too violent as it was. The crater, sixty feet across and thirty feet deep, became an obstacle to the Union advance. To cross it men had to climb out of a steep pit with no better foothold than loose sand. The poorly coordinated attack bogged down near the rim of the great hole. The Confederates wheeled up guns to blast at the men in the pit.

As the summer day dawned, one of the most gruesome scenes of the war played itself out. At times the crater became so crowded with Union troops that the dead had no room to fall but stood upright in the milling mass. The Southerners dropped mortar shells onto the heads of the attackers.

Then came a Confederate counterattack. "This day was the jubilee of fiends in human shape," a Southern soldier said, "and without souls." Having fired their weapons, the soldiers on both sides were reduced to fighting with the bayonet and the butts of their rifles. The battle became a savage reenactment of a medieval melee. When it was over the Union had lost more than 2,000 men, twice as many as the rebels.

"It was the saddest affair I have ever witnessed in the war," Grant wrote later. He was not sentimental about the loss of life, only about the wasted opportunity to break through the Confederate lines.

———

IN GUNPOWDER'S earliest days, many had seen it as an invention of the devil. Their views had been ratified many times over as men used the explosive to create hell on earth. Among the cataclysms that epitomized the horrors of the nineteenth-century gunpowder battle was the fight that took place at Gettysburg, Pennsylvania, over the first three days of July 1863. General Robert E. Lee carried the hopes of the Confederacy into Northern territory, seeking a decisive victory. The two armies met by accident outside the rural college town when the Confederates sent an expedition to seek some badly needed shoes. The first day, the Federals fell back onto the long swell of Cemetery Ridge south of the village, their line anchored on hills at both ends. The second day, Lee threw his forces into a coordinated attack at several points along this line, and nearly succeeded in breaking through. The third day, the two armies sat glaring at each other across a mile of open fields. A skirmish on the Union right died out at midmorning and a profound quiet settled across the landscape. For an hour or more, the only enemy that either side had to contend with was the mounting summer heat.

But Lee was not finished. He had mauled the Federals two days running, he could not simply retreat. If he was able to break the Union line with one final thrust, perhaps the glorious Cause of the Confederacy would be won. This was not true and Lee had enough strategic sense to realize it, but the word was passed to the Southern troops. Victory was at hand. All that was needed was courage.

Lee began with an artillery barrage. The impact of artillery in the middle of the 1800s was a direct extension of what Edward III had tried to achieve at Crécy five hundred years earlier: shock. The sound of the guns remained a critical element of their potency. The concussion could stagger advancing infantrymen even when the projectiles did not hit them. A 12-pounder cannon firing two and a half pounds of high-quality gunpowder produced a sensation that every man within range felt deep in his entrails. A cannonball could kill only a limited number of men, but the authoritative, knee-weakening blast of the guns could incite whole companies to run to the rear, spreading panic as they went. A Confederate artillery officer stated that he favored round shot, even at close range, because it was "more efficacious for breaking a charging line than shrapnel and canister, which while disabling twice as many did not make such a crashing noise."

At Gettysburg, the ten-man crews operated their guns in much the same manner as gunners had through the centuries, ramming powder and projectile down the barrel, and setting flame to the propellant through a narrow vent hole in the breech. Chemists had invented a primer that allowed the crew to fire the cannon by jerking a lanyard.

Emerging from the barrel at a speed of 1,200 miles per hour, a solid cannonball would still be flying faster than the speed of sound when it struck an object half a mile away. When fired on a flat trajectory, the shot hit the ground about 400 yards out, then ricocheted, remaining lethal as it bounced hundreds of yards farther. Each bounce sprayed debris that could itself kill or seriously injure nearby troops.

Alternatively, an artillery captain could choose to fire canister shells, thin metal cans that held 85 large lead bullets. The container blew apart as it exited the muzzle, allowing the shot to fly out in a cone pattern. At close range, the hail of metal was ruinous against massed troops.

Exploding shells had been lofted from mortars since gunpowder's earliest days. By the Civil War era, improved designs allowed them to be shot from cannon as well. Each 12-pounder cast iron shell was packed with six ounces of explosive gunpowder. When firing one, the gunners first estimated the range and adjusted the fuse, which would be touched off by the flame of the propelling powder. They tried to time the shell to explode over its target, spraying men, horses, and guns with fragments of iron.

In short, field artillery had become so lethal that it threatened attacking infantry soldiers or cavalrymen with devastation. Lee knew that it would be impossible to advance into the face of the Union artillery lined up on Cemetery Ridge. His plan was to drive off or disable enough guns to give his men a chance when they pressed forward. To do this, he brought together as many of his own guns as he could muster, 135 cannon in all, the greatest concentration of Confederate artillery ever assembled.

A sky of blue enamel looked down on the vast field. The air was shimmering with the July heat. At precisely 1:07 P.M.—a punctilious civilian mathematics teacher recorded the time—a two-shot signal broke the stillness. The desperate attempt was under way.

"As suddenly as an organ strikes up in church," a Confederate artillery officer recorded, "the grand roar followed from all the guns."

The firing of a single cannon makes a formidable sound that reverberates for miles. The roar of the Confederate battery, soon joined by a response from the Union guns, created an "infernal pandemonium" that is difficult to imagine. With the guns blasting hundreds of rounds a minute at the height of the bombardment, one soldier reported he "could distinguish no particular sound, it was one continuous and awful roar. The ground seemed to rock." Gunners were soon bleeding from both ears as a result of the concussions. Some soldiers were still stone deaf two days after the battle.

Smoke, the most distinctive feature of every gunpowder battlefield, immediately began to obscure the scene. It stung eyes and seared lungs. Gunners fired in virtual blindness. The "dull and lazy air," one witness recorded, "was now turned to a dark, wild and sulphurous atmosphere." The acrid taste of the powder filled the soldiers' mouths. "The sun through the smoke looked like a giant red ball."

Soldiers on both sides, waiting near their respective guns, clinging to cover where they could find it, endured an inhuman ordeal. Major Walter A. Van Rensselear of the 80th New York observed a "constantly moving arch of iron missiles screeching like fiends their defiance while passing each other in mid air." Shells approached with a quick hiss. Trees splintered. Dirt sprayed from bounding balls. One man felt the pressure of a shell passing overhead, looked back, and saw that it had "ploughed through the bodies of two men of my company . . . cutting their bodies literally in two."

The soldiers could watch shells exploding against the sky above them. "The flash was a bright gleam of lightning radiating from a point," a veteran said, "giving place in a thousandth part of a second, to a small, white, puffy cloud, like fleece of the lightest, whitest wool." Each of these explosions sprayed iron shell fragments in all directions. Color Sgt. John Dunn of the 1st Delaware said it was "like some horrible night-mare where one was held spell-bound by the appalling grandeur of the storm."

Yet for all the sound and fury, the Confederate bombardment weakened neither the capability nor the resolve of Union forces as much as Lee had hoped. Federal guns in the center were damaged— the ammunition limbers of many blew up, wreaking further destruction around them. But many of the Confederate rounds went high and landed in the rear of the Union battle line. A Prussian captain who attended the battle as an observer summed up the tremendous cannonade as *"eine Pulververschwendung"*—a waste of powder.

A natural instinct in the face of such a gunpowder storm was to dig. Yet the notion that such behavior was unworthy of a brave man still survived. At Gettysburg, Erasmus Williams of the 14th Virginia, forced to remain in place under the Yankee response to Lee's cannonade, began to dig a shallow pit using knife and bayonet. His lieutenant reprimanded him, "Why Williams, you are a coward."

"You may call me what you please," Williams replied, "but when the time comes I will show up all right."

The lieutenant proclaimed his own willingness to stand up and take whatever came. Early in the barrage he was smashed by a cannonball. His blood, Williams reported, "sprinkled all over me."

——

UNION COMMANDERS at Gettysburg, concerned that gunners were using up their ammunition, sent an order to cease firing. Union gun crews left off their feverish work around 2:30. The rebel guns continued their uproar for another half hour or so, until they too fell silent. The second part of Lee's plan was about to begin.

Leading the attack would be two division commanders, one the most intellectually accomplished officer on the field, the other a notorious dullard. Brigadier General James J. Pettigrew had graduated with highest honors from the University of North Carolina, spoke six languages, had traveled across Europe, and had written a book. Major General George E. Pickett had graduated last in his class at West Point and had compiled an undistinguished army record. His shoulder-length curls dressed in perfumed oil, he led 5,800 Virginians, the only fresh Confederate division on the field.

The Confederate troops, advised of the plan to attack the Yankee center, knew it was a desperate act. Some became "as still and thoughtful as Quakers at a love feast." One yelled out: "This is going to be a heller! Prepare for the worst!" Twelve thousand men stood up and organized themselves into lines, each more than a mile long. They dressed their ranks. They rested their rifles on their shoulders—there would be no shooting until they came up to the Union line. They stepped off.

The Civil War was the first conflict fought almost entirely with rifles, and it was the rifle that made this an affair of unprecedented gore. Most Confederate soldiers shouldered the crack Enfield; Union troops waited for them with Springfields, an American copy of the British model. Both rifles fired a conical Minié bullet just over half an inch across. Each mass-produced firearm cost less than $15.

Rifles extended both the range and accuracy of an infantrymen's fire. The Minié ball was accurate to at least a quarter mile. Federal soldiers, instead of throwing a volley of unaimed bullets, prepared to pick out targets with a reasonable expectation of hitting their man. The rifle was the defender's friend. From cover behind a tree or rock he could kill one attacker after another. Any advancing Southerner who stopped to take aim, fire, and reload became an easy target.

The accuracy and range of rifles were a lethal answer to the Napoleonic techniques that many Civil War generals had learned at West Point. Those tactics involved aggressive charges of massed troops intended to shock and overwhelm defenders until they either broke to the rear or were impaled on the bayonet. Most of the history of gunpowder firearms was represented on the field that day. Men of the 12th New Jersey hefted old smoothbore muskets. Except for its percussion primer, this gun was merely a variation on the arquebus of the fifteenth century. Two companies of a Connecticut regiment were armed with breech-loading Sharps rifles, which embodied many of the features of twentieth-century weapons. These guns could fire so fast that they overheated and the men had to pour water over the barrels.

As the Confederate line advanced, Union cannoneers fired solid shot directly into the masses of soldiers. They timed exploding shells to go off in front of the rows of troops. A well-placed shell could take out ten men at a time. Pettigrew's troops on the Confederate left were punished the worst by this artillery fire. Some panicked and either ran to the rear, surrendered, or simply lay down.

The majority of the Confederates came on. The Yankees who waited for them agreed that "it was a magnificent sight," a "grand soldierly effort." The thousands of men stretching across the field marched as if in a grand parade, regimental bands marking time, bayonets forming a river of steel. One Union private thought they gave "an appearance of being fearfully irresistible." That the Confederates stopped to dress their ranks even as the artillery shells were tearing into them amazed the Yankees.

The Northern soldiers, crouching or lying behind a low stone wall, were outnumbered on their immediate front by two to one. They held their fire while the rebel line approached. In earlier conflicts officers had simply tried to get their men to bring off a coordinated volley. Now their instructions were different. "Aim low," was the word passed along the line—in battle the men had a tendency to fire high. "Take careful aim." "Low and steady."

When the Confederate line came within range, the Union troops opened up and the fight began in earnest. The sound was intense and unnerving, "a continuous rattle like that which a boy makes running a stick along a picket fence, only vastly louder." For the rebels, the sudden fusillade was a "sleet storm, and made one gasp for breath." The advancing men leaned forward as if into a wind. Bullets rattled against a wooden fence, which formed the last obstacle before the Union line. Pettigrew's men "dropped from that fence as if swept by a gigantic sickle."

Only in the last thirty yards did the attackers break into a charge. By this time Union soldiers were shooting at them from their exposed flanks as well as from the front. A few rebels smashed into the Yankee lines. The fight reached a quick climax. "It was over in no time," a Union soldier remembered. The Federals rallied, held. Barely twenty minutes after it had begun, the greatest charge of the gunpowder era was finished. Surviving Confederates retreated to their lines, leaving behind more than half their number either dead, wounded, or captured.

A Union cannon and gun crew that opposed Pickett's charge at Gettysburg

"We gained nothing but glory" was how a Virginia captain summed up the experience.

That night it rained. Hundreds of injured soldiers lay in the no-man's-land between the two armies, moaning in pain, dying. Neither force was in a mood to renew the contest the next day. On the night of July 4th Lee retreated, carrying his injured soldiers on the agonizing journey back to Virginia.

After the war, Dr. Reed Bontecou, the director of an army hospital in Washington, made a photographic record of the injuries suffered by soldiers in the conflict. The images were eerily reminiscent of those *cartes de visite* that the optimistic warriors had prepared before they marched off to the fight four years earlier. All that had changed was that stumps had replaced arms and legs. All that had changed was that the uniforms were pulled aside to expose ulcerated wounds, misshapen torsos, grotesque deformities, the grievous effects of gunpowder weapons on unprotected flesh. All that had changed was that the proud faces had acquired a stunned and weary sadness.

13

THE OLD ARTICLE

"I WAS LOOKING up at the window over the door when, before I heard the noise, I recollect distinctly seeing the whole glass fly out of the window, each pane apparently whole, then break and fall." Lammot du Pont was 3 years old as this scene unfolded before him. Growing up in the midst of the world's biggest gunpowder factory, he would witness other strange and violent events during his lifetime.

Lammot pursued an interest in chemistry at the University of Pennsylvania. Unaffected by the melancholy that afflicted his grandfather, the tall, lanky youth enjoyed slim cigars and clowning. When he graduated in 1849, he was tempted to try his luck in the California gold fields. Instead, he joined the firm and soon took over supervision of the refinery, preparing saltpeter for the mill. The "family chemist" established a laboratory and spent his spare time investigating the qualities and possibilities of gunpowder.

Potassium nitrate had long been recognized as the ideal ingredient to provide the oxygen that turned the combustion of carbon and sulfur into an explosive deflagration. Its drawback was cost. Whether it was obtained from the soils of India or leached from rotting manure, its gathering and processing were expensive.

Sodium nitrate worked just as well as the basis of gunpowder, and unlike its potassium cousin, it was dirt cheap. Enormous quantities were available in the coastal regions of Chile. Its serious disadvantage was that it readily absorbed moisture, rendering the powder damp.

Lammot investigated ways to use this form of nitrate for making serviceable powder. Trial and error, informed by his insight into chemistry, led him in 1857 to patent an "improvement in gunpowder." It was the only fundamental alteration to the traditional formula that proved practical over gunpowder's long history.

Having ground sodium nitrate, sulfur, and charcoal together in the usual way, he tumbled the resulting grains with graphite for a full twelve hours. The graphite formed a coating that dramatically reduced the tendency to pick up moisture.

One characteristic of the new gunpowder, known as "B" blasting powder, was remarkable: It was of no use in guns. For the first time in the history of this ancient explosive, a variety had been formulated that was intended only for commercial purposes. These uses—blasting tunnels, working mines, leveling grades—had outpaced military applications.

His uncle Henry wisely promoted Lammot to partner in the firm and sent him to Europe in 1858 to make a thorough investigation of the state of gunpowder manufacturing across the Atlantic.

"From what I have seen of the mills here they are far behind us," Lammot reported. In spite of this confidence, Lammot paid careful attention to everything he saw as he made his grand gunpowder tour. At the venerable Waltham Abbey works in England, he investigated the

Lammot Du Pont (1831–1884)

modern steam engine that powered the rolling mills. The impeccable mill in Spandau, west of Berlin, boasted carpeting on all the floors— the works had not experienced an accident in more than twenty years. Visiting the French factories where his grandfather had studied with Lavoisier, Lammot saw stamp mills still in operation. The French, he was told, were working on a way to form nitric acid and combine it with potash in order to make saltpeter synthetically.

Back home, Lammot's blasting powder was proving a great success. One of the fastest-growing markets for powder was the anthracite coal mines of eastern Pennsylvania. The Du Ponts sold their product there through agents in mining towns like Mauch Chunk, Scranton, and Pittston. They shipped the powder first by wagon and canal, later by rail. But shipping costs, delays, and damage to the powder en route put the company at a disadvantage in relation to local mills.

In 1859, Henry du Pont, with Lammot's support, purchased a small, bankrupt gunpowder mill on the Wapwallopen Creek, twenty miles south of Wilkes-Barre. The move was commonplace on its surface. For the Du Ponts it was unprecedented. During more than half a century, they had produced powder only on the Brandywine. Close personal supervision by family members gave their products the highest reputation for quality. This small expansion marked a turning point in the industry as a whole. The Du Pont appetite for acquiring competing companies would prove insatiable. Over the next half century, the firm would aim at nothing less than a total monopoly over the making of gunpowder in America.

The Civil War left in its wake a glut of gunpowder and a surplus of gunpowder manufacturers. Entrepreneurs had set up and expanded mills to feed the appetite of the guns and the demands of civilian users. With peace, the government craving for gunpowder disappeared overnight. The market became crowded and hotly competitive. The small mills, many still barely past the craft stage of production, fought for local markets in mining areas. In the western United States, a group of miners had invested $100,000 to start the California Powder Works when their powder supplies were cut off during the war. This and other firms battled the eastern companies trying to reestablish a foothold in the West.

With the chaos in the industry, it was difficult for anyone to make money grinding powder in the postwar era. The solution larger firms hit on was market regulation: fixing prices and controlling competition. Monopoly was in the air in the late 1800s—Rockefeller sought to corner the market in oil, Carnegie in steel, Swift in beef. Even in this climate, the consolidation of the gunpowder industry was remarkable for its ruthlessness.

Already in his fifties as the war ended, Henry du Pont was the conservative, even cantankerous autocrat of the growing Du Pont empire. Notoriously parsimonious, he gathered willow branches as he walked the grounds and instructed that they be added to the supply of char-

A worker at the Du Pont plant beside kegs of gunpowder

coal. He resisted electricity and the typewriter, writing his letters with a quill pen by candlelight. Always a shrewd businessman, he had guided the partnership to solid profitability.

Charles Belin, a Du Pont in-law who ran their outpost on the Wapwallopen, noted in an 1870 letter that "financial matters are like military operations, the bigger force runs the day, the small fry will have to clear the track or go to the devil." Lammot took the same attitude. He headed a memo on the subject "Propose war." The first target of this war was the Schuylkill coal district around Pottstown, Pennsylvania. Miners blasted seven million tons of anthracite there every year. It was a buyer's market for gunpowder as both large and small powdermen vied for business.

As a bridgehead, Lammot secretly purchased the mill of Henry Weldy, leaving Weldy as front man. He then met with Albert Rand, president of Laflin & Rand, the New York–based gunpowder maker that was second only to the Du Pont firm. The two companies forged an agreement to divide the business in the Schuylkill fields. Their treatment of the "small fry" was to be strictly cutthroat.

First, the small mill owners were given a chance to sell out or join the Big Two in a cartel. Few agreed. Next, Du Pont dropped prices, hoping to force the "enemy mills" out of business. The campaign was intended, as Weldy put it, to "sicken the small men." It took longer than Lammot had hoped. Coal companies, the ultimate victims of the war, saw the direction events were taking and continued to buy from local producers, even at higher prices, to stave off the monopoly.

Solomon Turck, who became president of Laflin & Rand in 1873, questioned some of the rapacious tactics being proposed by his partner in the venture. "I feel that this world is big enough for all," he said, "and I know that we can take nothing out when we leave." The Du Ponts were adamant. Rights to water power were purchased in order to deprive competing mills. Workers were hired away from the enemy. Weldy intercepted the mail of his skilled employees to look for job offers. Machinery makers were threatened, mine bosses bribed. In the end the bigger force did run the day. By 1878 almost all the small mill owners had either sold out or gone bankrupt.

Henry had meanwhile expanded the war to the national level. In April 1872, joined by representatives from Laflin & Rand and a handful of other major gunpowder producers, he oversaw the formation of the Gunpowder Trade Association, which became known simply as the Powder Trust. The owners of the Warren Powder Company in Maine soon learned what it meant to defy the Trust. When they refused to keep their prices in line, Lammot sent a two-man delegation to demand compliance. They refused to be cowed. The cartel emissaries offered to buy them out. No go. The Association began to undersell Warren in its market area. The powdermen at Warren fought back, struggling on for seven years before finally descending into bankruptcy.

It didn't take too many incidents like this to put a healthy fear of Trust power into the minds of those whose money was invested in gunpowder mills. They generally preferred acquiescence to financial ruin. When the Trust was formed, the members already controlled 78 percent of the gunpowder market in the nation, with Du Pont responsible for 37 percent itself. These figures climbed as the consolidation proceeded. The day of the small gunpowder mill was over.

Even as the Trust was concentrating the industry, Du Pont was itself expanding. In 1876 the Delaware firm purchased the Hazard Powder Company, the third largest powdermaker in America—the change of ownership was kept secret. By the end of the 1870s, with Laflin & Rand the only serious competitor left standing, Lammot admitted that the Powder Trust was "only another name for Du Pont and Co."

By 1889, the Trust controlled 95 percent of the U.S. market for rifle powder and sold 90 percent of the blasting powder. Henry explained his perspective on the consolidation this way: "We do not allow anybody to dictate to us as to what price, terms, and conditions we shall dictate. We do our own dictating."

———

IN THE 1870s LAMMOT DU PONT RETURNED to the issue of improving gunpowder. As he knew from his collaboration with Thomas Rodman, who died in 1871, large guns needed a powder that burned slowly, generating pressure as the projectile moved up the bore. Lammot conducted hundreds of experiments to achieve a consistent powder providing the most favorable ballistics. At one point he meticulously counted the grains in several 100-pound barrels of powder to investigate the effect of grain size variation.

The result of these investigations, perfected in 1875, was a form of powder called Hexagonal. Shaped plates were used in the pressing to produce nuggets an inch and a half in diameter with a small hole in the middle—they were roughly the shape of wagon wheel nuts. Because the hole widened as the powder burned, the generation of hot gas increased while the space behind the projectile grew. More consistent than Mammoth powder, Hexagonal would be used in all large U.S. cannon for the next two decades.

Besides bringing gunpowder close to its most effective form, the collaboration between Rodman and Du Pont had represented one of the earliest science-based research projects for military purposes. In the 1850s rules of thumb and intuition still prevailed in the formulation of gunpowder. By the 1870s systematic research rooted in theory, mathematics, and accurate instruments was standard.

For their part, British powdermakers developed a highly compressed "rifle large-grain" powder for use in their larger artillery. In 1882, German technicians retarded burning still further by reducing the proportion of sulfur to only 2 percent and using charcoal made from rye straw that had not been thoroughly charred. They pressed this powder into inch-and-a-half-wide prisms with a central hole, a shape similar to Du Pont's Hexagonal powder. Brown prismatic powder, or "cocoa powder," topped all other powders in large guns: It threw projectiles the farthest with the most consistent trajectory and least strain on the gun. Artillerymen used it in massive 100-ton guns, which required a charge of more than 800 pounds of powder to blast a projectile out the barrel.

These were to be the last significant improvements in gunpowder. Over its long history, the mixture's possibilities had been thoroughly explored, its limitations starkly outlined. The next big developments, in both shooting and blasting, would be the discovery of substitutes for the ancient explosive. Gunpowder was about to become obsolete.

————

CHRISTIAN FRIEDRICH Schönbein, a professor of chemistry at the University of Basel, was an unsophisticated, roly-poly man who loved "eating sauerkraut, black sausages and dumplings." In 1840 he discovered ozone. In 1845 he dipped cotton fibers into a fuming combination of nitric and sulfuric acids. When he dried them, he found that they had become highly flammable, even explosive.

Gunpowder makers had long coaxed carbon fuel to interact with oxygen-rich nitrate by forcing charcoal and saltpeter into an intimate mixture. Yet the ingredients had always remained distinct chemicals. Schönbein had, in effect, incorporated the nitrate and carbon into a single molecule. The foundation was the cellulose of the cotton, a string of simple sugars that was the most common of all organic chemicals. The new material, nitrocellulose, or "guncotton," was unstable. A great deal of chemical energy was required to hold it together. When heat or a shock upset the balance, it broke apart, its material converting instantly to gas and releasing the binding energy as heat.

Schönbein knew immediately that he was onto something—he had stumbled across a potential rival to gunpowder. "Explosive cotton," he wrote, "should rapidly find a place in the pyrotechnic arts." He gave a demonstration of his discovery at Britain's Woolwich Arsenal in 1846. He presented Queen Victoria and the Prince Consort with the first brace of partridges killed with shot propelled by guncotton. Schönbein's hopes for turning the substance into a new form of gunpowder literally went up in smoke when the compound proved too volatile to be manufactured commercially.

The same year that Schönbein was presenting his game birds to the royal couple, an Italian chemist, Ascanio Sobrero, was trying the nitrification trick on glycerin, a by-product of soapmaking. Previous attempts had resulted in a noxious red vapor. But Sobrero carefully dripped the sweet-tasting syrup into a chilled bath of nitric and sulfuric acids, stirring the while. He created an insoluble oil that sank to the bottom of the container. He drew it off and washed it in water to remove the acids. This material, when struck sharply with a hammer, set off a window-rattling explosion. He called it "piroglicerina." Under the name nitroglycerin, it was destined to become gunpowder's first serious rival.

Sobrero was a man of scruples. "Science," he said, "should not be made a pretext or means of dishonorable undertakings or of business speculations." Not everyone agreed.

———

ONE MAN FAILED to be deterred by either the danger or the technical challenges involved in turning nitroglycerin into a money-making product. Alfred Nobel had been drawn into the explosives industry by his father Immanuel, a hustling Swedish entrepreneur who, after an early bankruptcy, had surfaced in St. Petersburg, Russia, in the 1840s. There he constructed some of the first floating gunpowder torpedoes ever to damage ships. He amassed a small fortune during the Crimean War, gave his son a first-class education, then lost everything and returned to Sweden. Searching for a better explosive with which to charge his mines, he turned to Sobrero's discovery.

Alfred was a sickly, morose youth who spoke five languages and loved the romantic poets, especially Shelley. But it was explosives that set fire to his imagination. During his late twenties, a determination of almost superhuman intensity overtook him. Even after his laboratory on the outskirts of Stockholm blew up, killing his 20-year-old brother Emil, he pushed forward. When nitroglycerin manufacture was banned inside the city limits, he moved his operation to a barge in a lake.

The problem that Nobel faced was how to get nitroglycerin to explode reliably. Unlike gunpowder, the oil did not go off when lit with a flame or fuse. Nobel's idea, remarkable in its simplicity, was to put a vial of gunpowder inside a container of nitro. Fire from a fuse set off the powder. The shock of the small explosion acted as a hammer to shake apart the nitroglycerin molecules, resulting in an energy-releasing chain reaction.

This simple idea of using the force of one explosive to detonate another opened the door to the use of synthetic chemical explosives. Nobel packed his initiator into a small copper container with an attached fuse, substituting mercury fulminate for the original gunpowder. He patented this blasting cap in 1864. The era of high explosives had arrived.

A gunpowder explosion differed from natural fire in the speed at which energy was released. The combustion of gunpowder generated the same amount of energy as ordinary burning, only much faster, bursting into hot gas and smoky residue in a few thousandths of a second. This lightning speed, though, was snail-paced compared to the explosion of nitroglycerin, which went off in millionths of a second. Imagine that the explosion of a gunpowder bomb were slowed down, so that half an hour elapsed between the touch of flame to the powder and the bursting of the container. On that same scale, a nitroglycerin explosion would be over in two seconds.

The reaction, known as a detonation, happened so suddenly that air molecules had no time to move out of the way of the expanding gases, but piled up on each other, forming a shock wave. The shock wave traveled outward faster than the speed of sound. Nitroglycerin needed no container to bring about an explosion, as gunpowder did.

The volatile new explosive got off to a shaky start. On November 15, 1865, a group of Sunday morning tipplers in the bar of the Wyoming Hotel, on New York City's Greenwich Street, complained of an odd smell drifting through the tavern. A porter carried a fuming box outside and set it down in the gutter. A few seconds later, a massive explosion shook the street, blowing out windows for a hundred yards. The parcel had belonged to a peddler hawking "glonoin oil," a new blasting agent. In San Francisco, in Panama, in Sydney, Australia, and Hamburg, Germany, accidental nitroglycerine explosions were doing even more damage. Alarmed governments took action. France and Belgium outlawed the possession of nitroglycerin. Great Britain excluded it. The U.S. Congress considered sentencing irresponsible shippers to hang. The age of high explosives seemed on the verge of ending before it had begun.

Nobel had almost single-handedly established the nitroglycerin industry, promoting its use around the world, setting up factories to make it. In light of these devastating explosions, he never blinked. Returning to his laboratory, he pursued the problem with his usual resolve. He suspected that many of the accidents that had given blasting oil a bad name had resulted from the liquid being spilled. He searched for a way to make spills impossible. After trying many absorbents—charcoal, sawdust, cement—he hit on diatomaceous earth, the tiny silica skeletons of algae, as the ideal dope. Kieselguhr, as it was called, absorbed three times its weight in nitroglycerine, and it turned the volatile chemical into a much more stable commodity. Like many important advances in technology, the idea was brilliant in its simplicity and seemed obvious after the fact. Even the name that Nobel selected for his creation was pure genius. He borrowed a word from the Greek that meant power, calling the nitroglycerin sticks "dynamite."

He patented this discovery in 1867. Now gunpowder faced a serious, versatile challenger as a blasting agent. To say that the dynamite industry's growth was explosive is neither pun nor exaggeration. Nobel packed the kieselguhr–nitroglycerin mixture into paper tubes and set out to sell it to the world. Buyers stood in line to purchase it.

Gunpowder manufacturers viewed the advent of dynamite with concern. Modern chemistry was, in essence, pulling the market out from under the powder business. As is so often the case in such circumstances, a pair of blinders proved more comfortable than a bold examination of the facts. Powdermen were encouraged by the early nitroglycerin accidents. Even after the introduction of dynamite, Henry du Pont declared that high explosives were "all vastly more dangerous than gunpowder, and no man's life is safe who uses them."

During the 1870s gunpowder makers watched their customers switch to high explosives in droves. The California Powder Company, of which the Du Pont family owned a controlling interest, began to feel the pressure first. Hard-rock miners of the West more and more preferred the dynamite being sold by the Giant Powder Company, which had licensed the rights from Nobel. Henry allowed the western firm to produce a competing product, but would not countenance it at the Brandywine works.

Many gunpowder makers remained loyal to the centuries-old product, which by now was being designated "black powder." Linus Austin of the Austin Powder Company in Cleveland still swore by "the old article," and figured customers would always prefer the "more conservative and more reliable black powder." Paul Oliver, one of the few anthracite powdermen whose firm, the Luzerne Powder Company, had not been gobbled up by the Du Pont cartel, considered high explosives "wretched stuff." "Decent and respectable black powder," he insisted, "will do more work for a given amount of dollars and cents."

But as the 1880s wore on, the challenge of high explosives mounted. Their promoters slapped ever cockier names on the variations that streamed out of the laboratories: Ajax, Rend Rock, Vigorite, Rippite, Earthquake Powder—there seemed no end to the parade of improvements.

As a chemist, Lammot du Pont understood that science was creating ever more effective substitutes for the product on which his family's empire rested. For a time, Uncle Henry swore by the "old article" as fiercely as any other powderman. But in 1880 he gave Lammot permission to set up a dynamite factory. "We are going into the high explosives business," the old man announced.

Lammot built a modern dynamite works from scratch in Repauno, New Jersey, carefully separating the buildings with earthen berms. One Saturday in 1884 he was called to see about "trouble in the nitro house." He found a vat containing a ton of fuming, unstable nitroglycerin. He ordered the men out, began tapping the oil into a pool of water, and left the building himself. At that instant, the nitro exploded, killing him. The manufacture of explosives had not ceased to be a dangerous business.

The dynamite industry grew faster than any other business in history, hammering the lid on the coffin of gunpowder. Nobel became fabulously wealthy. In 1888 he opened a newspaper and read his own obituary. His brother Ludvig had died recently and the reporter mistook the name. Alfred was stunned to find that he was described as a "Merchant of Death."

To assure that his actual obituary would read more favorably than the mistaken one, Nobel wrote a will leaving the bulk of his fortune to establish prizes for those who made notable contributions to the sciences, literature, and world peace. In a final ironic twist, Nobel developed angina pectoris in his last years. His doctor prescribed nitroglycerin, which had become the standard remedy for the heart ailment.

———

IN THE 1860s, the Albany, New York, company Phelan & Collender, the nation's leading manufacturer of billiard equipment, offered a tempting $10,000 prize to anyone who could come up with a substitute for ivory, long used to make billiard balls but increasingly difficult to come by. A man named John Wesley Hyatt took up the challenge. Though a printer by trade, he was an avid inventor and knew of the experiments that had been done with nitrated organic chemicals.

One of those chemicals was a sticky, syrupy form of nitrated cellulose called collodion, originally developed by Schönbein, which had been used to dress wounds during the Civil War. Hyatt dissolved this chemical in camphor under high heat and pressure. The result was a material that could be molded when hot—it was "plastic." When cooled, it became as hard as the ivory he was trying to replace. Hyatt called it celluloid. It was the first of the long parade of synthetic materials that we now know as plastics.

Munitions researchers took a keen interest in celluloid. Guncotton and nitroglycerin were far too explosive to use in guns. The chemicals detonated rather than burned, blowing up the chamber instead of propelling the bullet. The process for making celluloid, based on a raw material similar to the explosive nitrogen compounds, suggested to them a way to unseat gunpowder from its role as a propellant.

The demands on a propellant were many and subtle compared to those on an explosive. The propellant had to ignite reliably and burn evenly. It had to be able to withstand shock, and couldn't absorb moisture or generate noxious fumes when fired. Ideally it would not produce smoke or residues. That the craftsmen who developed gunpowder over its long history were able to make natural ingredients fulfill many of these requirements is testimony both to their skill and to the almost magical versatility of the mixture itself. Chemists struggled to produce an equivalent.

What was wanted was a hard material, energetic but resistant to detonation, that could be fashioned into grains. The material would be very similar, in fact, to the plastic that Hyatt had developed for billiard balls. During the 1880s, French inventor Paul Vieille became the first to find a way to plasticize guncotton by mixing energetic nitrocellulose with a solvent of ether and alcohol. The dried result was a hard, nonporous material that he could form into any size particles. When it burned, it created little smoke or residue. Vieille unveiled this "smokeless" powder, known as *Poudre B,* in 1886.

Alfred Nobel developed a rival propellant two years later by including nitroglycerin in the mix. Alarmed at the progress on the Continent, the British formulated a brew similar to Nobel's, adding petroleum jelly. The mixture formed a dough that could be forced through dies to produce long strings—the new propellant was christened "Cordite."

One prominent advantage of smokeless powder, as its name implies, was that on burning virtually all of its material turned to hot gas. The combustion products of gunpowder were half gaseous, half solid. Besides engendering smoke and fouling the inside of the barrel, the solid material reduced the force of the explosion. Smokeless powder was considerably more powerful than the same weight of gunpowder.

The new synthetics, in a remarkably short time, displaced gunpowder from its 900-year-old niche as the world's only effective propellant. By the 1890s nations across Europe were rushing to switch to the improved powder. At first, smokeless powders needed to be distinguished from standard gunpowder. But the balance quickly shifted.

"Gunpowder" began to refer to the modern synthetic, whereas "black powder" was used to designate the ancient mixture.

———

JUST AS EUROPEANS had taken advantage of their early lead in gunpowder weapons to support a wave of world conquest, they ushered out the age of powder with a second and more comprehensive program of colonial domination. Their superiority in mass production and metalworking, which allowed them to use gunpowder with increasing effectiveness, made the difference. From 1617 until 1852, British authorities issued 300 patents connected with firearms. During the six years ending in 1858, inventors took out 600 more such patents. This frantic pace of arms development accelerated through the latter part of the century.

Native blacksmiths in Africa and Asia had been able to copy or repair muskets quite handily. But they lacked machine tools and sophisticated metalworking capability to reproduce or even fix modern rifles. Repeating rifles in particular gave small groups of Europeans such a superiority of firepower that they could face down vastly larger numbers. The Chinese still used heavy, old-fashioned cannon that Jesuits had cast for Ming emperors 300 years earlier. The Burmese required each soldier to grind his own gunpowder. British conquest of these areas was child's play.

The last and easiest target was Africa. So sure were European rulers of their military superiority that they divided the continent into spheres of influence even before venturing out to subdue the natives. They had good reason for confidence. They had long been shipping muskets to Africa as part of the slave trade, but the weapons were of consistently poor quality. Locally ground gunpowder was also inferior. The British used Gatling guns against Zulus during wars in 1871 and 1879. That same decade, General Garnet Wolseley defeated the mighty West African Ashanti kingdom with a force of only 6,500 soldiers. The French put the machine gun to use in Egypt in 1884. They fired repeating rifles to bring down Senegalese fighters armed with spears and poisoned arrows. Many Africans in the interior had no experience with firearms at all. "The whites did not seize their enemy as we do by the body, but thundered from afar," reported one survivor of an encounter with European forces. "Death raged everywhere—like the death vomited forth from the tempest."

Americans also dabbled in imperialism, and in doing so had the distinction of taking part in the last major conflict that used gunpowder fashioned from saltpeter, charcoal, and sulfur. In March of 1898, prompted by a rabble-rousing press and by the mysterious explosion of the battleship *Maine* in Havana harbor, U.S. president William McKinley declared war on Spain. Neither the U.S. military—the army had only 28,000 regular troops—nor the Du Pont partnership, which by now effectively controlled the American gunpowder market, were ready for the hostilities.

Efforts to develop effective smokeless military powder in the United States had languished. Many American soldiers carried the Model 1889 firearm, known as the "trapdoor rifle" because of its rather primitive breech-loading mechanism. It used cartridges loaded with black powder. The trapdoor rifle suffered from a number of disadvantages compared to the smokeless, repeater rifles used by Spanish troops. The worst was the cloud of smoke that announced the shooter's position to enemy riflemen, inevitably leading to additional American casualties.

In spite of the disadvantages of their reliance on black powder, U.S. troops won the Spanish-American War handily by August. Afterward, American armed forces quickly converted to rifles and artillery firing smokeless powder. The military use of the original gunpowder, whose tortuous history stretched back to the raids of Edward III and to the battles of medieval China, had come to an end.

———

IN 1875, ALFRED Nobel, who was helping bring about the demise of gunpowder, wrote something of an epitaph for the ancient substance: "That old mixture possesses a truly admirable elasticity which permits its adaptation to purposes of the most varied nature. Thus, in the mine it is wanted to blast without propelling; in a gun to propel without blasting. . . . But like a servant of all work, it lacks perfection in each department, and modern science, armed with better tools, is gradually encroaching on its old domain."

At the dawn of the twentieth century, that encroachment was almost complete. For military purposes, black powder was obsolete, replaced by smokeless powder and high explosives. Suppliers would continue to sell black powder cartridges for older sporting guns. Even in the 1920s some sportsmen swore by black powder as equal, if not superior, to synthetic propellants, especially in shotguns.

High explosives manufacturers claimed that their product "surpasses the power of gunpowder as much as the ball of a gun surpasses in swiftness and destructive power the dart of an Indian." In spite of this acknowledged power, using high explosives underground was tricky business, and it was in the mines that gunpowder hung on the longest. Coal miners in particular were reluctant to switch to newer explosives. They maintained an affection for gunpowder blasting long after modern chemistry began offering safer and more effective explosives. Gunpowder had been tested over centuries; it was familiar, predictable, and inexpensive. Miners used it regularly up until the 1920s, giving it up only reluctantly. More than a few stuck with the old article into the 1950s, when gunpowder finally disappeared from the mines.

In response to this continued viability, the Du Ponts, in 1888, twenty years after the introduction of dynamite, began building what would become the largest black powder plant in the world. The site was near Keokuk, Iowa. The works, known as the Mooar Mills, was the first Du Pont operation west of the Mississippi. In addition to being a shrewd effort to profit from the growing Midwestern coal fields, the mills were Henry du Pont's final expression of faith in the product that had made his fortune—he died a year after they opened. The firm expanded the mills in 1892, again in 1900, and one last time in 1918. At the operation's peak, sixteen pairs of grinding wheels were turning out an astonishing 80 tons of powder every day.

In 1902 the Du Pont firm celebrated its centennial by swallowing its last significant competitor in the gunpowder business, Laflin & Rand. Five years later, President Teddy Roosevelt turned his trust-busting sights on the powder monopoly. A federal court dismantled the firm, splitting off the Atlas Powder Company and the Hercules Powder Company in an attempt to restore competition in the explosives market.

In 1921, a year after a last accidental explosion, the company shut down its gunpowder mills on the Brandywine. The grinding wheels had been turning there continually for 117 years. In 1971, Du Pont announced that it was leaving the black powder business altogether.

Black powder still found uses throughout the twentieth century. Quarrymen working with slate and similar rock needed the "soft" explosion that was characteristic of gunpowder, not the shattering effect of high explosives. Fuse makers still used gunpowder because of its predictable burn rate. It served as a primer for large navy guns, setting off the smokeless powder that did the heavy lifting. The ranks of diehard sports shooters who preferred the old powder were joined by black powder enthusiasts who enjoyed the challenge of hunting with old-fashioned weapons. Hobbyists and history buffs took pleasure in reenacting bygone battles, forming another small market for traditional powder. Their games reproduced the smoke and smell of bygone battlefields.

And so, step by step, gunpowder eased off the world stage much as it had eased on, without fanfare. No headlines trumpeted the demise of a 900-year-old technology. No eulogy was read over its grave. Gunpowder was gradually replaced, fading into the haze of nostalgia. In only one field did its contribution continue intact. Appropriately, it was the same use to which the ancient Chinese had first put the magical and incendiary substance—delight.

———

THE HEART OF a modern fireworks display is a cardboard sphere six inches across. Inside is a small charge of gunpowder. The device is in fact a bomb and is often referred to as such by professionals. Its purpose is not destruction but enchantment. It contains nuggets of chemicals that, when set afire by the exploding gunpowder, fly through the air, burning with brilliant sapphire, crimson, or golden light. No matter how much modern apparatus accumulates on the periphery of his trade, the modern pyrotechnician continues to ply his art with gunpowder tools that are almost identical to those that have decorated the sky and ignited the imagination for centuries.

Attached to the sphere, also known as an aerial shell, is a small sack of gunpowder, the lift charge. The "shooter" inserts the whole device into a three-foot-long tube and ignites a fuse that quickly burns down to set off the lift charge. The explosion blows the sphere 800 feet into the air. An attached timing fuse, lit by the initial firing, ignites the powder inside the shell just as it reaches its apogee. The burst flings the flaming nuggets out against the black velvet night.

The pyrotechnician takes gunpowder back to its origins. He reenacts mankind's primordial impulse to use the substance to captivate and gladden his fellows. Like any showman, he devotes hours to the prosaic preparations that lead to thirty minutes of spectacle. He endures the inevitable danger always associated with gunpowder. The electric fuses that fire some shows have reduced the risk, but accidents, sometimes fatal ones, continue to happen.

Fireworks remain, in most aspects, a product of craft. The shells, the fountains, the fiery chrysanthemums and peonies, all are made by hand. A show may be programmed by a computer, but the ingredients, the planning, the salutes and whistles and skyrockets, would be familiar to the pyrotechnicians who illuminated Louis XIV's Versailles. During the nineteenth century, more sophisticated chemistry allowed pyrotechnicians to create dramatic new effects. More varied

A fireworks show near the Washington Monument

and richer colors became available as the salts of various metals were made to glow in the flames. Finely powdered magnesium added a brilliant white light. At the opening of New York's Brooklyn Bridge in 1883, more than a million spectators gazed in awe at the riverwide

"Niagara of Fire," a cascade of magnesium flame that spilled from the new span.

Today, fireworks professionals and a few devoted amateurs continue a tradition that reaches back to the Renaissance and beyond. The pyrotechnician is more than a showman. He is heir to the alchemist, transforming the basest of materials—a handful of charcoal, some sulfur, salt leached from ordure—into gold. With that gold he etches the darkness.

A fireworks display gives the audience a symbolic reprise of gunpowder's history—the concussions of bombards and cannon stretching back to Crécy, the explosions that carved the earth, the blossoms of fiery stars that enchanted monarchs and peasants centuries ago, the long pageant of joy and terror. The haze of smoke that lingers after the show brings to modern nostrils the smell that perfumed the night air in medieval China.

Many have tried to describe the evanescent beauty of fireworks. The explosions are splendid waste. They are wild-haired comets, silver rain, tinsel-starred bouquets. In 1540, Vannoccio Biringuccio wrote that fireworks "endure no longer than the kiss of a lover for his lady, if as long." This ephemeral quality is the deepest secret of their appeal. However spectacular the incandescence, it fades in a moment. The quick and heartbreaking diminuendo instills fireworks with a bittersweet poignancy that has drawn involuntary sighs of wonder from onlookers down the ages.

EPILOGUE:
SOMETHING NEW

ON JULY 16, 1945, a small article appeared in the *Albuquerque Tribune:* "An ammunition magazine, containing high-explosives and pyrotechnics, exploded early today in a remote area of the Alamogordo air base reservation, producing a brilliant flash and blast, which were reported to have been observed as far away as Gallup, 235 miles northwest."

The story was a lie. At 5:30 that morning a group of scientists had initiated a new era in human history.

The saga had begun fifty years earlier. In 1895 the German physicist Wilhelm Röntgen discovered a mysterious energy emanating from the glass walls of a cathode-ray tube. When he covered the tube with black paper, the rays still lit up a screen of fluorescing material. He passed his hand in front of the tube. A faint shadow formed on the screen. Within it, in darker outline, he could clearly discern the silhouette of his bones.

Röntgen's report on what he called X-rays stunned the world of science. It suggested an entirely new direction for research — the investigation of a novel, undreamed-of form of energy. Discoveries came quickly. Searching for X-rays, the Frenchman Henri Becquerel found a similar form of energy radiating from uranium. In 1898 Marie Curie and her husband Pierre isolated the element radium, a much more potent source of what she called radioactivity.

At McGill University in Montreal a lanky New Zealand farm boy turned physicist named Ernest Rutherford teamed up with the English chemist Frederick Soddy to look into a curious phenomenon that Rutherford had noticed during his own research into radioactivity. He had detected an unusual "emanation" from the radioactive element thorium. Soddy analyzed the gas and found it to be a different element altogether—argon.

Rutherford and Soddy posited an impossibility. An atom, by definition, could not be divided. Yet with radioactive emissions, elements were altering into entirely new materials by giving up part of their substance.

"Rutherford, this is transmutation!" Soddy exclaimed.

"Don't call it transmutation," Rutherford shot back. "They'll have our heads off as alchemists."

Transmutation it was, the metamorphosis of matter that, over a span of two millennia, countless alchemists and amateurs had devoted lifetimes to seeking. Just as astounding, the two scientists asserted in a 1903 paper, was the amount of energy involved in the process, which had to be "at least twenty-thousand times, and may be a million times, as great as the energy of any molecular change."

We can only imagine the emotions felt by the Chinese alchemists who created the first chemical explosion on earth. Perhaps they were similar to Soddy's feelings.

"I was overwhelmed with something greater than joy," he said. "I cannot very well express it—a kind of exaltation."

Soddy speculated that the energy contained inside the atom, if it could be tapped, could one day "make the whole world one smiling Garden of Eden." He also envisioned a darker side to the discovery. Back in England, lecturing to the Corps of Royal Engineers, he discussed the possibility that someone could develop a "weapon by which he could destroy the earth."

In another of the ironic echoes that are the grace notes of history, radioactivity during the early twentieth century assumed a role as an elixir of life reminiscent of the potions prescribed for the emperors of medieval China. Physicians raced to try out cures based on the amazing new energy. They were encouraged when they found that it reduced some skin cancers and that the waters of many of the world's famous spas were mildly radioactive. Was it possible that the invisible rays stimulated the body toward health? "Old Age May be Stayed by Radium" a newspaper proclaimed. Radioactive patent medicines proliferated—they proved as toxic as the elixirs of the alchemists.

It remained for a man of imagination to limn the true nature of the new form of energy. In 1914, H. G. Wells published a prophetic novel entitled *The World Set Free*. "The history of mankind," it began, "is the history of the attainment of external power." A quarter century before scientists split the atom, Wells foresaw a form of "atomic disintegration" that would unleash limitless power. The result would be no smiling Eden, but cities shattered by the "unquenchable crimson conflagrations of the atomic bombs."

As dawn broke on that summer day in 1945, there appeared on earth something whose arrival was truly commensurate with the advent of gunpowder. The blast, referred to by the code name "Trinity," produced light stronger than any that had ever shone on earth, heat 10,000 times the temperature of the sun's surface, a force 20 million times greater than that of a high explosive. Like gunpowder, the atomic bomb would have a profound and unexpected impact on the nature of warfare. Like gunpowder, it would prove difficult to tame for peaceful uses and would elicit references to supernatural forces. Like gunpowder, it would bring a new form of terror into the world. Like gunpowder, it would awe its very creators.

"Naturally, we were very jubilant over the outcome of the experiment," remembered Isidor Rabi, one of the scientists who watched the bomb explode. "Then, there was a chill, which was not the morning cold."

SOURCES

Because this book is intended as a popular account of gunpowder, not a scholarly work, I have forgone the use of footnotes to identify references. The following notes will indicate the sources of some of my research and will serve as a guide to those who wish to delve more deeply into the subject.

GENERAL

Brenda Buchanan at the University of Bath has been a leader in advancing the study of gunpowder technology and manufacturing. She is the editor of a collection of scholarly essays entitled *Gunpowder: History of an International Technology* (Moorland Publishing, 1996), which provides authoritative and wide-ranging resources on gunpowder's rich history. A second volume is in the works.

George I. Brown's *The Big Bang, A History of Explosives* (Sutton Publishing, 1998) offers an interesting and clearly written overview of explosives from gunpowder to nuclear fusion. In *The Chemistry of Powder and Explosives* (Angriff Press, 1972), MIT chemistry professor Tenney Davis gives details about all things explosive. First published in 1941, the book ranges from fireworks formulas to the chemical characteristics of nitrosoguanadine. Oscar Guttmann was one of the leading authorities on explosives at the end of the nineteenth century. His book *The Manufacture of Explosives* (Whittaker & Co., 1909) is full of valuable details about gunpowder.

Arnold Pacey traces the complex relationship of culture and technology in *Technology in World Civilization: A Thousand-Year History* (M.I.T. Press, 1990). Alfred W. Crosby's book *Throwing Fire: Projectile Technology Through History* (Cambridge University Press, 2002) is a succinct and elegant summation of the subject. John Keegan's *A History of Warfare* (Knopf, 1993) and William H. McNeill's *The Pursuit of Power* (University of Chicago, 1982) are two excellent and thought-provoking works that discuss the broader context within which gunpowder developed.

The former Du Pont gunpowder factory in Wilmington, Delaware, has been turned into the Hagley Museum. The mills and workshops still stand, and the library contains a voluminous collection of materials dealing with gunpowder history. In England, the gunpowder works at Waltham Abbey, Essex, are maintained as an educational and tourist attraction.

CHAPTER 1—FIRE DRUG

Joseph Needham's *Science and Civilization in China* (Cambridge University Press, 1986) is the most comprehensive source of information about gunpowder in ancient China. Needham spent a long lifetime documenting Chinese technical accomplishments. His scholarship is staggering; his sleeve-clutching insistence when it comes to Chinese precedence is difficult to resist. Volume 5, Part 7, of the encyclopedic work is the one relevant to "fire drug."

Imperial China 900–1800 by Frederick W. Mote (Harvard University Press, 1999) gives an excellent overview, particularly of China's relationship with the nomads of inner Asia.

CHAPTER 2—THUNDRING NOYSE

Modern scholarship on gunpowder's history began in 1960 with *A History of Greek Fire and Gunpowder* (Johns Hopkins University Press reprint, 1998), by the chemist and science historian James R. Partington. Professor Partington's terse style and his failure to provide translations of Greek and Latin quotation can be off-putting to casual readers, but his work remains among the most important in the field.

Philippe Contamine's *War in the Middle Ages* (Blackwell, 1984) gives one of the most authoritative pictures of the military world in which gunpowder arose.

Barbara Tuchman's remarkable account of the fourteenth century, *A Distant Mirror* (Knopf, 1979), provides a rich flavor of that time as well as details about Edward's campaigns.

CHAPTER 3— THE MOST PERNICIOUS ARTS

In *Joan of Arc: A Military Leader,* (Sutton Publishing, 1999) eminent military historian Kelly DeVries gives a wonderful analysis of Joan's real accomplishments. Professor DeVries' *Medieval Military Technology* (Broadview Press, 1992) is a meticulously researched exploration of early methods of fighting.

David Nicolle's *Constantinople 1453*, (Osprey Publishing 2000) is a clear and nicely illustrated overview of the momentous battle. Franz Babinger provides more detailed information about the victor at Constantinople in *Mehmed the Conqueror and His Time* (Princeton University Press, 1978).

CHAPTER 4—THE DEVILLS BIRDS

An intelligent discussion of the early days of gunpowder and gunpowder weaponry in Europe can be found in *Weapons and Warfare in Renaissance Europe* (Johns Hopkins University Press, 1997) by Bert S. Hall. Professor Hall's work combines groundbreaking scholarship with a highly readable style, a welcome combination in the field of technical literature.

Vannoccio Biringuccio's *Pyrotechnia* was reprinted by The M.I.T. Press in 1966. The book is a fascinating artifact of the sixteenth century. The author's last chapter deals with "the many sublimates and smoky tinctures" of "the burning and most powerful fire of love," a touching coda to a grim subject.

Carolo M. Cipolla's *Guns, Sails, and Empires* (Pantheon, 1966) is a lucid study of the nitty-gritty of early cannon and their impact on the world. In *Giving Up the Gun* (D. R. Godine, 1979) Noel Perrin tells the fascinating story of the rise and decline of gunpowder in Japan. *Guns: An Illustrated History of Artillery* (New York Graphic Society, 1971), edited by Joseph Jobé, is another useful guide.

CHAPTER 5—VILLAINOUS SALTPETRE

A History of Fireworks (George G. Harrap & Co., 1949) by Alan St. H. Brock is a seminal source of information about the early development of pyrotechnics. Brock's family operated one of England's leading fireworks companies. George Plimpton's more recent *Fireworks: A History and Celebration* (Doubleday, 1984) is a quirky, lavishly illustrated treatment of the subject.

Simon Pepper's *Firearms and Fortifications* (University of Chicago Press, 1986) describes the ways in which forts and gunpowder weapons evolved together. *Theater of Fire* (Society for Theatre Research, 1998) by Phillip Butterworth is an entertaining volume that delves into the gunpowder effects used in early theatrical performances. A discussion by J. R. Hale entitled "Gunpowder and the Renaissance: An Essay in the History of Ideas" contains a wealth of information about how gunpowder was perceived during its early days. It can be found in Charles H. Carter's *From the Renaissance to the Counter-Reformation* (Random House, 1965).

CHAPTER 6—CONQUEST'S CRIMSON WING

Two outstanding books give details of gunpowder in naval warfare. Peter Padfield's *Guns at Sea* (St. Martins, 1974) is beautifully illustrated and reliable. *The Arming and Fitting of English Ships of War 1600–1815* (Conway Maritime Press, 1987) by Brian Lavery digs even more deeply into maritime minutia.

In *Gunpowder and Galleys* (Cambridge University Press, 1964) John F. Guilmartin gives an excellent overview of the evolution of war at sea and the role of gunpowder in the process. *Sea Life in Nelson's Time* was written in 1905 by British Poet Laureate John Masefield. Reissued by the Naval Institute Press in 2002, it contains sharply etched images of naval life.

CHAPTER 7—NITRO-AERIAL SPIRIT

Wayne Cocroft gives a very detailed and meticulously researched account of early gunpowder manufacture in *Dangerous Energy* (English Heritage, 2000). In *Essays and Papers on the History of Modern Science* (John Hopkins University Press, 1977), Henry Guerlac includes a chapter called "The Poet's Nitre: Studies in the Chemistry of John Mayow," which provides interesting details about Mayow's theories as they relate to gunpowder.

Cecil J. Schneer's *Mind and Matter* (Grove Press, 1969) and John Read's *Through Alchemy to Chemistry* (Harper and Row, 1957) both offer valuable insights into the early days of chemistry. Joseph Needham explored the idea of the gunpowder origins of the internal combustion engine in *Gunpowder as the Fourth Power* (Hong Kong University Press, 1983).

CHAPTER 8—NO ONE REASONS

One of the most readable of the many works on the Powder Treason is Antonia Fraser's elegant *Faith and Reason* (Doubleday, 1996). Fraser's understanding of the times and personalities that formed the context of the plot contribute to a clear picture of an affair that remains murky in some respects even today.

Michael Roberts's book *Gustavus Adolphus and the Rise of Sweden* (English Universities Press, 1973) untangles many of the complicated strands of the Thirty Years' War. *A History of Arms* (AB Nordbok, 1976) by William Reid, is a very clearly illustrated account of the entire history of weapons and gives interesting details about this period.

CHAPTER 9—WHAT VICTORY COSTS

Two works that offer important information about the earliest developments in ballistics are *Ballistics in the Seventeenth Century* (Cambridge University Press, 1952) by A. R. Hall; and *Firepower: Weapons Effectiveness on the Battlefield 1630–1850* (Scribner, 1975) by B. P. Hughes. Jenny West's *Gunpowder, Government, and War in the Mid-eighteenth Century* (Boydell & Brewer, 1991) contains much original data about the manufactured and handling of gunpowder in England.

Of Arms and Men: A History of War, Weapons, and Aggression (Oxford University Press, 1989) by Robert L. O'Connell is a wide-ranging and detailed discussion of the impact of weaponry on warfare. The author's insights about the stagnation of weapons development in the eighteenth century are especially illuminating.

CHAPTER 10—HISTORY OUT OF CONTROL

David H. Fischer's engrossing discussion of the outbreak of the American Revolution, *Paul Revere's Ride* (Oxford University Press, 1994), provides a vivid narrative that illuminates the role gunpowder played in the drama. Orlando W. Stephenson uncovered a good deal of information about the source of American gunpowder in the ensuing war and reported it in "The Supply of Gunpowder in 1776," an article printed in the *American Historical Revue* (Volume 30, 1925).

Jean Pierre Poirier's *Lavoisier: Chemist, Biologist, Economist* (University of Pennsylvania Press, 1997) is a thorough account of the French scientist's career. Robert Multhauf documented the efforts to produce gunpowder during and after Lavoisier's administration in "The French Crash Program for Saltpeter Production, 1776–94" which appeared in *Technology and Culture* (Volume 12, 1971).

CHAPTER 11—THE MEETING OF HEAVEN AND EARTH

William Carr's *The Du Ponts of Delaware* (Dodd, Mead, 1964) is a readable and balanced account of the gunpowder family. William Hosley produced an elegant, lavishly illustrated study called *Colt: The Making of an American Legend* (University of Massachusetts Press, 1996). Paul Wahl and Donald R. Toppel's *The Gatling Gun* (Arco Publishing, 1965) is the definitive work on its subject.

Lee M. Pearson's article "The 'Princeton' and the 'Peacemaker': A Study in Nineteenth-Century Naval Research and Development Procedures," which appeared in *Technology and Culture* (Volume VII, Number 2; 1966), brings out many details of the tragic incident. Daniel B. Webster, Jr., wrote a detailed description of Thomas Rodman's often underappreciated accomplishments in "Rodman's Great Guns," which appeared in *Ordnance* (July–August 1962).

CHAPTER 12—APPALLING GRANDEUR

Paddy Griffith's *Battle Tactics of the Civil War* (Yale University Press, 1987) discusses the weapons of the war and how they were used. *Pickett's Charge: The Last Attack at Gettysburg* (University of North Carolina Press, 2001) by Earl J. Hess is a riveting look at the greatest attack of the gunpowder age. "The Augusta Powder Works: The Confederacy's Manufacturing Triumph," by C. L. Bragg, M.D., which appeared in *Confederate Veteran* (Volume 1, 1997), discusses George Washington Rains' effort to supply his side during the war. Milton Perry's *Infernal Machines* (Louisiana State University Press, 1985) tells the story of the Confederacy's mine and submarine warfare.

CHAPTER 13—THE OLD ARTICLE

History of the Explosives Industry in America (reissued by Ayer Company Publishers, 1998) by Arthur Pine Van Gelder and Hugo Schlatter was originally published in 1927. The book, weighing in at 1,132 pages, gives not only an exhaustive picture of an industry, but also a flavor of the days of mutton-chopped magnates.

Norman B. Wilkinson's *Lammot du Pont and the American Explosives Industry* (University Press of Virginia, 1984) provides an informative discussion of Lammot's career, the Powder Trust, and the later days of the gunpowder industry. Herta Pauli's *Alfred Nobel: Dynamite King—Architect of Peace* (L. B. Fischer, 1942) is a balanced portrait of the originator of high explosives.

I want to acknowledge the many scholars who were generous with their time and information, and the librarians who went out of their way to help me track down obscure sources. Also: Jack Fielder, who provided a wealth of pyrotechnic detail; William Knight, who shared his insights and the results of his research into the technical aspects of gunpowder; Loretta Barrett, who backed the project enthusiastically from the beginning; and Joy Taylor, who read and commented on more than one draft of the book and provided moral support. Thanks to all.

ILLUSTRATIONS

The figures on p. 16 are taken from *The Fire-Drake Manual* (or Huo Lung Ching), dated circa 1412. The illustrations on pp. 35, 37, and 44 are derived from

German manuscripts dating from the fourteenth to the sixteenth centuries. The images on pp. 42, 60, 68, 75, and 137 are all taken from *The Gunner: The Making of Fire Works*, by Robert Norton, first published in 1628. The portrait of Berthold Schwartz on p. 24 is from *Pourtrait et vies de hommes illustres*, by André Thevet, 1584. The drawing on p. 26 is adapted from "Bellifortis," an illuminated manuscript dated circa 1400. The image on p. 58 is from the Munich Codex Germanicus #600. The drawing of a soldier on p. 67 is from *Art Militaire au Cheval*, by J.J. von Wallhausen, 1616. Page 72 shows a woodcut by Sebald Beham, dating back to the early sixteenth century. The fire-breathing dragon on p. 83 is from *Pyrotechnia*, by John Babington, 1635. The somewhat fanciful warship on p. 92 is from a sixteenth-century engraving by Pieter Brueghel the Elder.

The illustrations on pp. 46, 71, and 150 are courtesy of Joy Taylor Graphic Design; figures on pp. 163, 175, 177, 179, 219, and 221 all appear courtesy of the Hagley Museum and Library in Wilmington, Delaware; and the images on pp. 166, 215, and 237 are from the Library of Congress Prints & Photographs Division.

INDEX